ABSORPTION SPECTRA OF MINOR BASES THEIR NUCLEOSIDES, NUCLEOTIDES, AND SELECTED OLIGORIBONUCLEOTIDES

SPEKTRY POGLOSHCHENIYA MINORNYKH OSNOVANII IKH NUKLEOZIDOV, NUKLEOTIDOV I NEKOTORYKH OLIGORIBONUKLEOTIDOV

СПЕКТРЫ ПОГЛОЩЕНИЯ МИНОРНЫХ ОСНОВАНИЙ ИХ НУКЛЕОЗИДОВ, НУКЛЕОТИДОВ И НЕКОТОРЫХ ОЛИГОРИБОНУКЛЕОТИДОВ

ABSORPTION SPECTRA OF MINOR BASES

Their Nucleosides, Nucleotides, and Selected Oligoribonucleotides

Tat'yana Vladimirovna Venkstern
Aleksandr Aleksandrovich Baev

Authorized translation from the Russian

Springer Science+Business Media, LLC 1965

The Russian text was published in Moscow in 1965 by Nauka for the Institute
of Molecular Biology of the Academy of Sciences of the USSR.

Спектры поглощения минорных оснований, их
нуклеозидов, нуклеотидов и некоторых олигорибонуклеотидов

*Татьяна Владимировна Венкстерн
и Александр Александрович Баев*

Library of Congress Catalog Card Number 65-27344

©1965 Springer Science+Business Media New York
Originally published by Plenum Press Data Division in 1965.
Softcover reprint of the hardcover 1st edition 1965

ISBN 978-1-4899-6277-5   ISBN 978-1-4899-6602-5 (eBook)
DOI 10.1007/978-1-4899-6602-5

*No part of this publication may be reproduced in any
form without written permission from the publisher*

Table of Contents

Introduction 1

Abbreviations 13

Spectra and Tables 15

    1. Adenosine-3'-Phosphate 16

    2. Guanosine-3'-Phosphate 18

    3. Uridine-3'-Phosphate 20

    4. Cytidine-3'-Phosphate 22

    5. 1-Methyladenine 24

    6. N_6-Methyladenine 26

    7. N_6-Dimethyladenine 28

    8. 1-Methylguanine 30

    9. 1-Methylguanosine 32

    10. 1-Methylguanosine-3'-Phosphate 34

    11. N_2-Dimethylguanine 36

    12. N_2-Dimethylguanosine 38

    13. N_2-Dimethylguanosine-3'-Phosphate 40

    14. N_2-Methylguanine 42

    15. Pseudouridine 44

16. Pseudouridylic Acid. 46

17. Thymine Riboside 48

18. Thymine Riboside-3'-Phosphate. 50

19. 5-Methylcytosine. 52

20. 5-Hydroxymethylcytosine 54

21. Adenylyl-3',5'-Cytidine-3'-Phosphate. 56

22. Adenylyl-3',-5'-Uridine-3'-Phosphate. 58

23. Guanylyl-3',5'-Cytidine-3'-Phosphate. 60

24. Guanylyl-3',5'-Uridine-3'-Phosphate 62

25. Adenylyl-3',5'-Guanosine-3'-Phosphate. 64

26. N_2-Dimethylguanylyl-3',5'-Cytidine-3'-Phosphate. . 66

27. N_2-Dimethylguanylyl-3',5'-Pseudouridine-3'-Phosphate 68

28. Adenylyl-3',5'-Adenylyl-3',5'-Cytidine-3'-Phosphate 70

29. Adenylyl-3',5'-Adenylyl-3',5'-Uridine-3'-Phosphate 72

30. Guanylyl-3',5'-Guanylyl-3',5'-Uridine-3'-Phosphate 74

31. (Adenylyl-3',5'-Guanylyl-3',5')-Uridine-3'-Phosphate 76

32. (Adenylyl-3',5'-Guanylyl-3',5')-Cytidine-3'-Phosphate 78

33. 1-Methylguanylyl-3',5'-N_2-Methylguanylyl-3',5'-Cytidine-3'-Phosphate. 80

Literature Cited 83

ABSORPTION SPECTRA OF MINOR BASES

Their Nucleosides, Nucleotides,
and Selected Oligoribonucleotides

Introduction

GENERAL ASPECTS

The primary structure of the nucleic acids did not attract the attention of researchers until recently, and its study apparently did not meet with success. But the last four to five years have substantially changed the situation. Investigations, begun in the laboratories of Nirenberg and Ochoa, on the stimulation of protein synthesis in vitro by artificial polyribonucleotides evoked interest in the primary structure of the nucleic acids and promoted the development of experimental research along this line. Data on the transmission of information during protein synthesis have not only confirmed the hypothesis of the template nature of protein synthesis, but also have made it possible to formulate a rule of correspondence between the linear sequences of amino acids in proteins and the bases in the nucleic acids — the code of protein synthesis. In addition, it has become clear that the codons (in the form of triplets or doublets — in this case there is no difference) are not the only method of recording metabolic commands, the source of which is the nucleic acids. Actually, the code of protein synthesis, as it was formulated in the investigations mentioned above, serves to transmit information pertaining to protein synthesis; but in reality it acts only at the level of the nucleic acids. The limited significance of the code is detected most distinctly in the adaptor function of t-RNA, the anticodon of which participates only in its interaction with m-RNA, while the reaction with the amino acid, itself nonspecific since it is accomplished by an acceptor...CCA terminal group, common to all t-RNA, acquires the properties of selectivity only as a result of the fact that t-RNA forms a coenzyme-enzyme type complex with aminoacyl-t-RNA synthetase. The structure of the contact portion of t-RNA, which enters into a bond with the active center of the synthetase, is unknown, and there is no basis for assuming that it possesses anything in common with the structure of the codons. It may be assumed that in other cases of specific interaction of nucleic acids with proteins as well (possibly also with other compounds) there are analogous relationships. Thus, the Nirenberg-Ochoa code cannot be considered universal and the only possible system of structural organization of nucleic acids, although, of course, its significance in this regard should not be underestimated.

Investigations of the code of protein synthesis have served to indicate the modernity of the attack on the problem of the primary nucleic acid structure. Many years of experiments in the investigation of the secondary nucleic acid structure also clearly indicate the significance of the sequence of nucleotide residues for the macromolecular organization of the nucleic acids. The physicochemical trend in the study of the nucleic acids has also promoted a shift to investigation of their primary structure, which has now been extended both to DNA and to RNA, and, in particular, to the transfer ribonucleic acids.

The successes achieved in the last six to seven years in the study of the function and structure of t-RNA have been especially important for the development of research in the field of the primary nucleic acid structure. The low molecular weight of t-RNA (about 25,000) and the correspondingly small length of their polynucleotide chains, comprising an average of 70–80 nucleotide residues, permit us to hope that the primary structure of these nucleic acids will prove accessible to decipherment by presently existing facilities. A favorable prerequisite for research is the possibility of formulating the criteria of homogeneity of t-RNA, being guided at first by their functional properties, primarily by their specificity with respect to amino acids, and by the degeneracy of the codons. These criteria could not only be formulated, but also utilized in practice in the isolation of individual t-RNA's, which places the latter in an exceptional position in comparison with the other nucleic acids, for which the criteria of homogeneity at present are less distinct. Work on the isolation of homogeneous t-RNA's is now under way in many foreign (e.g., Soviet) laboratories; such studies have led to the production of high-purity t-RNA preparations and have given the first, although incomplete, data on their primary structures.

A pecularity of t-RNA structure lies in the presence not only of the four principal bases in the molecules, but also of the so-called minor components — pseudouridylic and ribothymidylic acids, nucleotides containing methylated derivatives of adenine, guanine, cytosine, uracil, and certain other compounds, nonuniformly distributed along individual polynucleotide chains and among individual t-RNA's. The minor components facilitate (or at least should facilitate) the task of analyzing the primary structure of t-RNA, and eliminate to a definite degree the difference in the number of structurally differing components which exists between the proteins, containing 20 different amino acids, and the nucleic acids, which contain four main bases. It is natural that all researchers who have studied the primary structure of the nucleic acids have paid attention to the minor components, the chemical and physicochemical properties of which have already been rather well studied, while their functional role remains obscure in many respects.

In a study of the primary structure of the nucleic acids, many procedural methods have been developed and tested; there is no need to outline them here, since in general they boil down to a degradation of the nucleic acid molecules by chemical or enzymatic agents, resulting in the production of monomer units (nucleotides), or blocks of them (oligonucleotides). The latter are again subjected to cleavage, but using different methods, until structural units accessible to direct identification are obtained: mononucleotides, dinucleotides, and in certain cases trinucleotides.

SPECTROPHOTOMETRY OF NITROGEN BASES AND ITS APPLICATION—BROMINATION

Spectrophotometry of nitrogen bases and their derivatives is the main method of identifying these compounds, completing the preceding stages of chemical and chromatographic analysis. There are a number of handbooks which cite the absorption spectra and basic spectrophotometric parameters of the principal nitrogen bases and their derivatives. Data pertaining to the absorption spectra of the minor components are scattered in various journal articles, and their use in daily work involves great difficulties. The spectra of the simplest oligonucleotides generally have not yet attracted the proper attention, although they may be successfully used for the identification of these compounds. In a first approximation, the spectra of the oligonucleotides are additive and possess the characteristic features of their components. Our study of the absorption spectra of the lower oligonucleotides at various pH values and after bromination has shown that, just as in the case of the nitrogen bases and their derivatives, they may be used for analytical purposes [1, 2].

Bromination has already long been used in the identification of nitrogen bases and their derivatives [3, 6, 8, 14, 20]. As is well known, the addition of bromine results in the disappearance of the absorption bands of guanine, uracil, and cytosine in the region of 250–280 mμ, while the spectrum of adenine in the range 220–300 mμ [15, 21] remains unchanged. This fact permits a reliable distinction of uracil and its derivatives from the corresponding compounds of the adenyl series, in spite of the similarity of their spectra [5]. Experiments have shown that the nitrogen bases contained in the oligonucleotides behave in the same way during bromination, and that the characteristic absorption bands of any oligonucleotide in the region of 250–280 mμ disappear only if the oligonucleotide does not contain adenylic acid or any of its methylated derivatives. Otherwise, the spectrum of the oligonucleotides is converted to the spectrum of adenylic acid or the corresponding methylated derivative. This circumstance, in conjunction with certain other, more secondary features of the spectra (for example, the appearance of a new small maximum at 290–300 mμ when cytosine is incompletely brominated or the high position of the residual curve in the case of guanine), makes bromination an extremely effective method in the identification of oligonucleotides.

An important advantage of the spectrophotometric method of identification is the saving of time and material. Spectra in acid and alkaline media and after bromination may be taken successively on one sample; this process requires only 20–30 μg of the substance.

CONTENTS OF THE ATLAS

The experience used in the compilation of this atlas was acquired in connection with a study of the primary structure of transfer RNA, and this has made a definite impression upon the materials cited in it. First of all, the atlas contains the spectra of all the minor components that were available to us. In certain cases, the spectra were obtained not only for the minor bases, but also for the corresponding nucleosides and nucleotides. Moreover, the atlas contains the spectra of the four main mononucleotides, as well as certain di- and trinucleotides, including those containing the minor components.

The publication of the atlas of spectra is aimed at providing researchers studying the primary nucleic acid structure with handbook material that can be used in routine work in the identification of minor components and oligonucleotides.

In our opinion, the atlas may be useful to anyone employed in the study of t-RNA. Actually, oligonucleotides consisting of the principal bases are formed in the cleavage of any ribonucleic acids; moreover, the absorption spectra of the deoxy-series differ from the absorption spectra of the corresponding ribonucleic derivatives only in details, and hence the spectra cited may also be useful in the analysis of DNA. Furthermore, it is known that methylated bases are encountered not only in t-RNA but also, of course in very negligible quantities, in the ribosomal and other species of RNA [9-12, 24], as well as in DNA [18, 22, 23]. Thus, the spectra of the minor components may be of use in the analysis of practically any nucleic acids.

RECORDING OF ABSORPTION SPECTRA

We should emphasize primarily that all the absorption spectra cited in the atlas were obtained under conditions of the usual analytical work, without the use of methods that would be inaccessible to the standard biochemical laboratory possessing any spectrometer (for example, the SF-4) at its disposal. Although the spectra cited in the atlas were taken on a recording spectrophotometer (Hitachi Company, EP2), analogous spectra may in principle be obtained on any spectrophotometer, by measuring the extinction of the solutions for individual wavelengths and then outlining the absorption curve according to the data obtained. In this case, the extinction may be measured at 5 mμ intervals over the entire spectrum, with the interval of measurement reduced to 2 mμ only at the maxima and at points of inflection.

The spectra cited were outlined in a linear wavelength scale, since in the literature the absorption curves are given precisely in this form. In the recording of the absorption spectra under different conditions (in acid and alkaline media and after bromination), special attention was paid to the adequacy of the controls.

PRÉPARATIONS

We had at our disposal four minor components in the form of crystalline preparations (5-methylcytosine, 5-hydroxymethylcytosine, 2-methylaminopurine from the Calbiochem Company, and 2-dimethylaminopurine from the California Corporation for Biochemical Research). The spectra of these compounds were taken after a preliminary verification of their purity by paper chromatography in three solvent systems (see below). The remaining minor components were isolated from alkaline hydrolyzates of t-RNA on Dowex-1 × 8, HCOO, according to Cantoni's method [7]; the compounds were identified, checked for homogeneity, and purified also by paper chromatography. Since t-RNA was cleaved by alkaline hydrolysis, we obtained a mixture of 2'- and 3'-phosphates, which we used for recording the spectra. As is well known, the spectra of mononucleotides are practically independent of the position of the phosphate residue; hence, we made no effort to record the spectra of these isomers separately. The structural formulas for the 3'-phosphates are arbitrarily cited in the corresponding tables.

The nucleosides were produced by the action of prostate phosphomonoesterase on the corresponding nucleotides. After dephosphorylation, the compounds were again checked for homogeneity and purified by paper chromatography.

The purine bases were produced by hydrolysis of the nucleotides in 1N HCl at 100°C for 1 hr, after which chromatographic purification was also conducted.

The oligonucleotides were isolated from the pyrimidyl-RNA-ase (ApGp from guanyl-RNA-ase) of the t-RNA hydrolyzate in columns with Dowex-1 × 4, HCOO⁻, then checked for homogeneity and purified by paper chromatography.

Eluates of chromatograms obtained with the most volatile solvent, giving the purest spectra, were always used to record the spectra (see below).

Verification of the homogeneity of the peaks obtained in separation by ion exchangers, as well as their supplementary partition in the case of inhomogeneity, was accomplished by the method of two-dimensional paper chromatography, using the solvent system isobutyric acid — 0.5 N NH_4OH (10:6), pH 3.7, in one direction [16], and the solvent system tertiary butanol — HCOOH, pH 3.8 (1:1), pH 4.8, in the second [17]. The compounds giving one spot each, i.e., those that proved homogeneous, were transferred to a new chromatogram, which was run in the solvent system ethanol — 1 M ammonium acetate buffer, pH 7.5 (70:30). This solvent guarantees the greatest purity of the background of the chromatograms and substances to be separated. All the spectra cited in the atlas were taken on substances purified using this solvent.

FORMULATION OF THE TABLES AND QUANTITATIVE DATA

For each compound, the spectrum in 0.1 N HCl (curve 1), that in 0.1 N KOH (curve 2), and that after bromination (curve 3) are cited. For the most part, two names are given, one of which corresponds to the rules of Geneva nomenclature, while the second is the common name or represents an abbreviation in accord with the rules used in the literature. All the tables present the structural formulas, gross formulas, and the molecular weights computed on the basis of them.

Furthermore, for each compound the positions of the absorption maximum and minimum in $m\mu$ are presented for solutions in 0.1 N HCl and 0.1 N KOH. We should mention that these values fluctuate from case to case within a range of 1–2 $m\mu$, and this should be considered when using them for identification. The positions of the absorption maxima and minima cited in the literature also differ; in the case, for example, of N_2-dimethylguanine in an alkaline medium, the differences exceed the permissible limits [13, 19].

The molar extinction coefficients ($\epsilon_{max} \times 10^{-3}$ and in certain cases $\epsilon_{min} \times 10^{-3}$) were taken from the literature in most cases, since we were unable to determine them because of a shortage of material. We determined the molar extinction coefficients only of 5-methylcytosine, 5-hydroxymethylcytosine, 6-methylaminopurine, and 6-dimethylaminopurine (these data are given in parentheses in the tables), as well as of the dinucleotides ApCp, GpCp, and $G^M p\psi Up$. There are definite discrepancies between our data and the literature data, just as there are among the data cited by different authors. The molar extinction coefficients of the dinucleotides calculated on the basis of a phosphorus determination coincided with the sum of the molar extinction coefficients of the component nucleotides.

The characteristic values for various compounds are ratios of the absorption at different wavelengths. Each table presents four ratios for acid solutions and four for alkaline solutions; in all cases the absorption at a given wavelength was related to the absorption at 260 mμ. As our experience has shown, these values also vary from experiment to experiment (especially in an alkaline medium), and deviations of hundredths should be considered normal. Nonetheless, the combination of these ratios is characteristic of various compounds, and considering possible fluctuations, they can be used successfully for identification. We shall not cite the extinction ratios and other data for brominated samples, since, as was stated earlier, the absorption bands of the oligonucleotides in the region 220–320 mμ lose their characteristic features as a result of bromination.

We should emphasize that the identification should be performed if possible on the basis of the combination of indices under different experimental conditions, keeping in mind the possible and permissible deviations. In such an approach the spectrophotometric identification of minor components in the lower oligonucleotides can be performed reliably enough, with but a small expenditure of time and material.

PRACTICAL SUGGESTIONS

In the isolation of minor components or oligonucleotides from mixtures, one should first of all make sure that the compound is actually homogeneous; this can be accomplished by two-dimensional paper chromatography in the systems described above. Then the substance from a spot that has proved homogeneous is purified either on Dowex-1, Cl$^-$ [4], or by paper chromatography. In the latter case, the substance is applied to a new chromatogram, which is developed with the solvent system ethanol — 1M ammonium acetate buffer, pH 7.5 (70:30). The spot, which should be as compact as possible, is eluted from such a chromatogram with 2–3 ml (depending on the amount of the substance) of 0.1 N HCl. It is also useful to verify the homogeneity of crystalline preparations by paper chromatography; in the case of contaminations, not a weighed sample of the substance, but the eluate of the corresponding spot on the chromatogram should be used to obtain the spectrum.

The optimum value of E of solutions in the obtaining of spectra is $E_{max} = 0.600-0.700$; work should be aimed at achieving this value.

The absorption spectrum of a solution of a purified compound in 0.1 N HCl is taken in comparison with the same acid (curve 1 — spectrum in acid medium). Then a 1/20 volume of 4N KOH is added to the sample, and the spectrum is taken against 0.1 N HCl alkalized in the same way (curve 2 — spectrum in alkaline medium). The sample is reacidified by adding a 1/20 volume of 4N HCl, and bromination is conducted by adding 0.1 ml saturated bromine water to 3 ml of the eluate. After 10 min has elapsed, the excess bromine water is removed by aeration (10–15 min) and the spectrum is recorded against a control prepared in exactly the same way (curve 3 — spectrum after bromination). The dilution due to the addition of alkali, acid, and bromine water is neglected.

The data obtained — the general outlines of the absorption curves in acid and alkaline media, the influence of bromine, the positions of the maxima and minima, the ratio of the extinctions at different wavelengths — are compared with the corresponding atlas data. The results of the spectrophotometric investigations should be compared as far as possible with data on the chromatographic mobility (one's own or literature data). An identification may be considered reliable only when all the data coincide, of course, considering permissible deviations (see above).

In conclusion, let us note certain facts, the causes of which have remained obscure:

1. As is well known, the spectra of nucleosides and nucleotides are very close, practically identical. Despite this, we noted differences between the spectra of 1-methylguanosine and 1-methylguanosine-3-phosphate, as well as between certain parameters of thymine riboside and TMP; in particular, upon alkalization, ϵ_{max} is the same for 1-methylguanosine at acid and alkaline pH, while for the corresponding nucleotide, ϵ_{max} in 0.1 N HCl < ϵ_{max} in 0.1 N KOH.

2. The greatest fluctuations in the ratios E_{250}/E_{260}, etc. are always noted for methylated guanine derivatives at an alkaline pH. This has also been observed by other authors.

We should like to express our deep gratitude to L. P. Shershneva, who was extremely helpful in the compilation and formulation of the atlas.

Abbreviations

t-RNA................Transfer (adaptor, soluble) RNA.
m-RNA...............Template (messenger) RNA.
Pyrimidyl-RNA-ase....Pancreatic RNA-ase, polyribonucleotide-2'-oligonucleotidotransferase (cyclizing) (KF 2.7.7.16).
Guanyl-RNA-ase.......RNA-ase T_1; 3'-guanylohydrolase of nucleic acids (KF 3.1.4.8). (The oligonucleotides are designated in accord with the rules proposed by Markham and Smith; the capital letter denotes the corresponding nucleoside; the letter p to the right the phosphate residue added to the 3'-OH group of ribose; the letter p standing to the left, the phosphate residue added to the 5'-OH group of ribose. For example, ApGp — adenylyl-3',5'-guanosine-3'-phosphate.)
λ....................Wavelengths in mμ.
E....................Extinction; $E = -\log_{10} T = 2 - \log_{10}(100T)$, where T is the transmission of the sample in a cuvette with optical cross section 1 cm, as compared to a solvent whose transmission is equal to 1.
ϵ....................Molar extinction coefficient (E/c, where c is the concentration in moles per liter).

Spectra and Tables

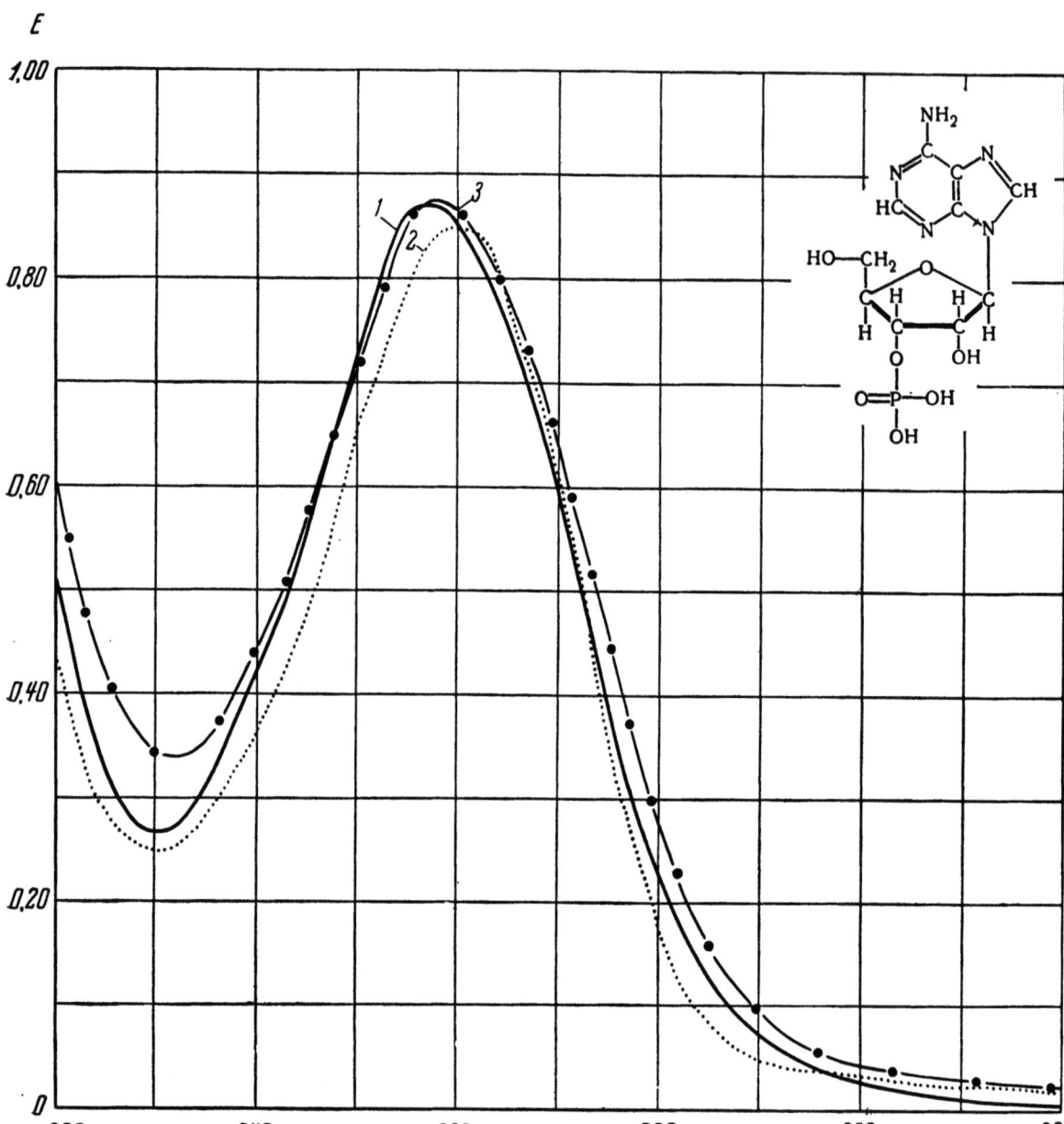

Table 1

ADENOSINE-3'-PHOSPHATE [AMP]

6-Aminopurineriboside-3'-phosphate

$C_{10}H_{14}N_5O_7P$

Mol. wt. 347

	1)	2)
	0.1 N HCl	0.1 N KOH
λ_{max} (mμ)	257	259
$\varepsilon_{max} \times 10^{-3}$	15.1	15.4
λ_{min} (mμ)	230	227
$\varepsilon_{min} \times 10^{-3}$	3.5	2.6
E_{250}/E_{260}	0.86	0.78
E_{270}/E_{260}	0.71	0.73
E_{280}/E_{260}	0.27	0.22
E_{290}/E_{260}	0.08	0.05

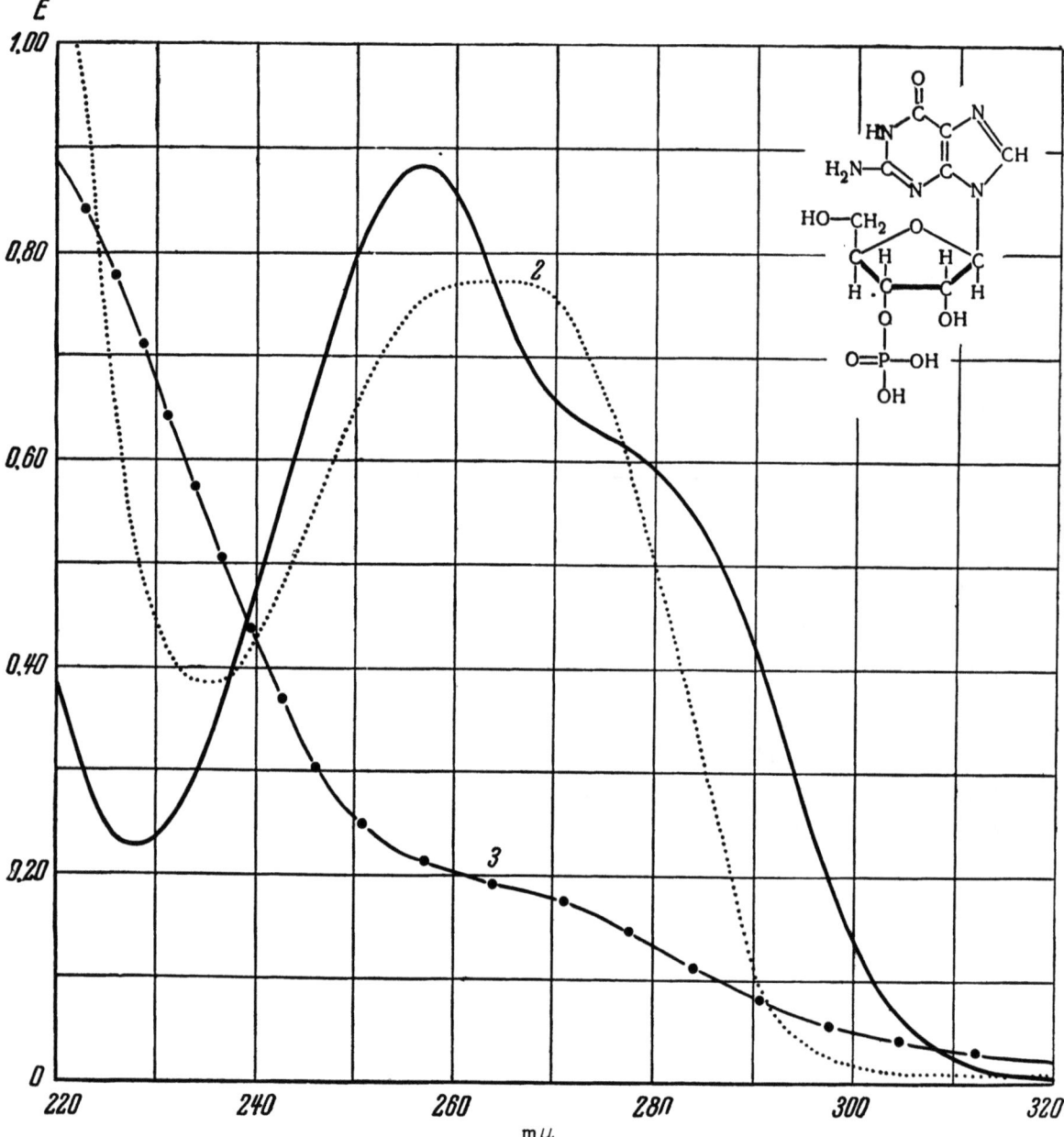

Table 2

GUANOSINE-3'-PHOSPHATE [GMP]

2-Amino-6-hydroxypurineriboside-3'-phosphate

$C_{10}H_{14}N_5O_8P$

Mol. wt. 363

	0.1 N HCl	0.1 N KOH
λ_{max} (mμ)	256	258
$\varepsilon_{max} \times 10^{-3}$	12.2	11.6
λ_{min} (mμ)	228	230
$\varepsilon_{min} \times 10^{-3}$	2.5	4.3
E_{250}/E_{260}	0.94	0.84
E_{270}/E_{260}	0.77	0.99
E_{280}/E_{260}	0.70	0.66
E_{290}/E_{260}	0.49	0.14

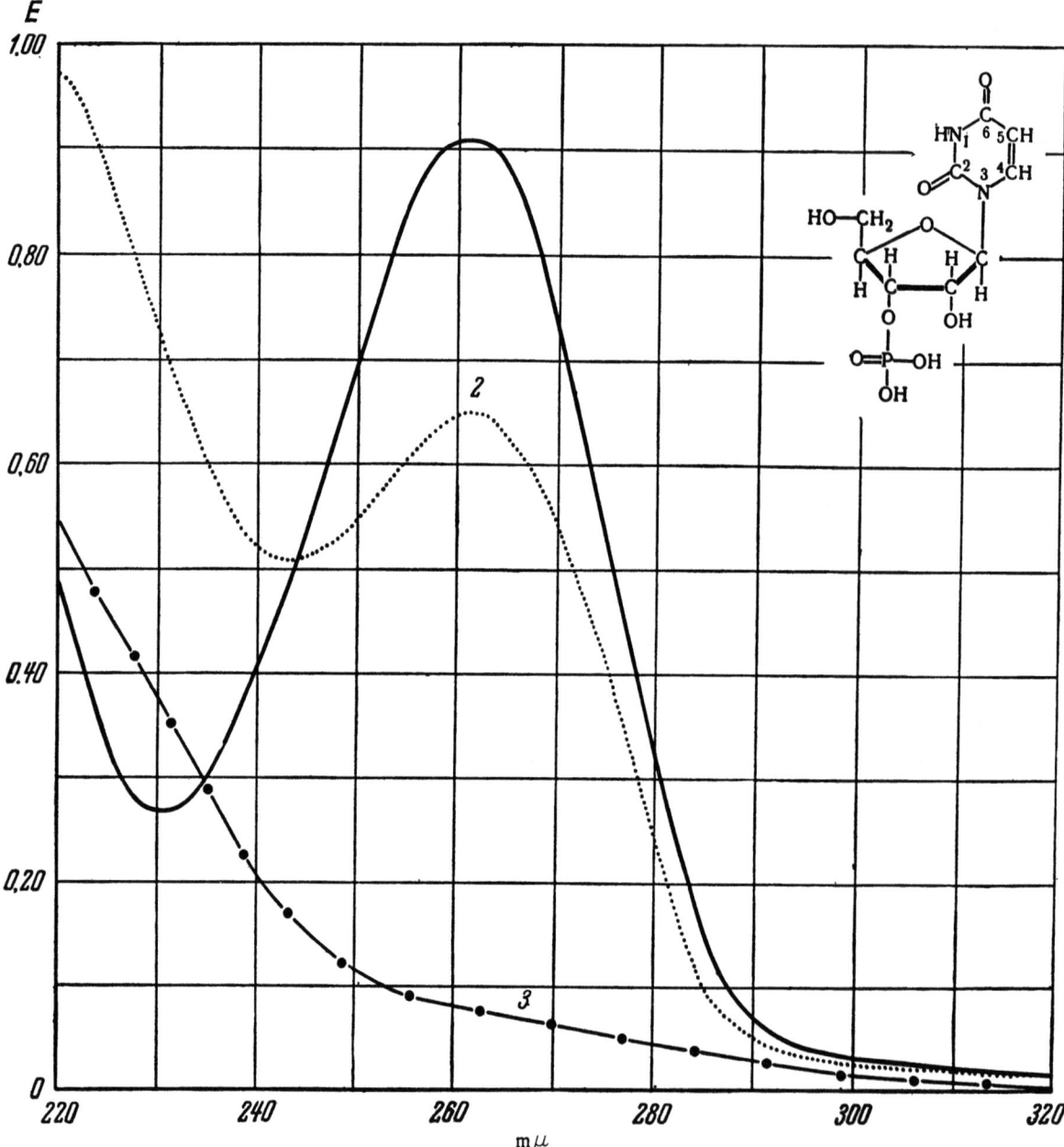

Table 3

URIDINE-3'-PHOSPHATE [UMP]
2,6-Dihydroxypyrimidineriboside-3'-phosphate
$C_9H_{13}N_2O_9P$
Mol. wt. 324

	0.1 N HCl	0.1 N KOH
λ_{max} (mμ)	262	261
$\varepsilon_{max} \times 10^{-3}$	10.0	7.8
λ_{min} (mμ)	230	241
$\varepsilon_{min} \times 10^{-3}$	2.1	5.3
E_{250}/E_{260}	0.77	0.85
E_{270}/E_{260}	0.82	0.85
E_{280}/E_{260}	0.37	0.37
E_{290}/E_{260}	0.07	0.08

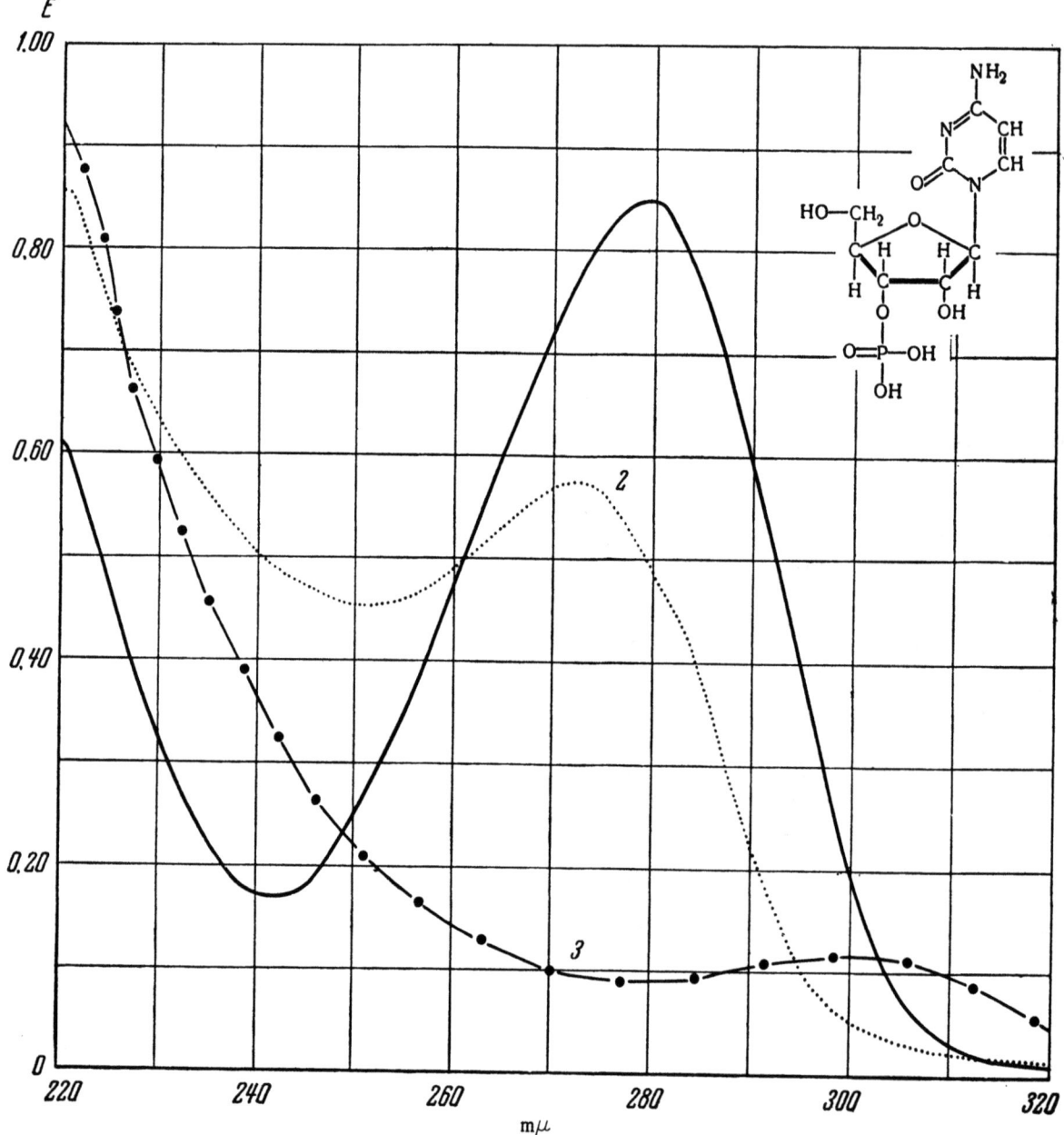

Table 4

CYTIDINE-3'-PHOSPHATE [CMP]
2-Amino-6-hydroxypyrimidineriboside-
3'-phosphate
$C_9H_{14}N_3O_8P$
Mol. wt. 323

	0.1 N HCl	0.1 N KOH
λ_{max} (mμ)	280	271
$\varepsilon_{max} \times 10^{-3}$	13.0	9.0
λ_{min} (mμ)	241	249
$\varepsilon_{min} \times 10^{-3}$	1.6	6.2
E_{250}/E_{260}	0.53	0.94
E_{270}/E_{260}	1.51	1.16
E_{280}/E_{260}	1.77	0.98
E_{290}/E_{260}	1.26	0.44

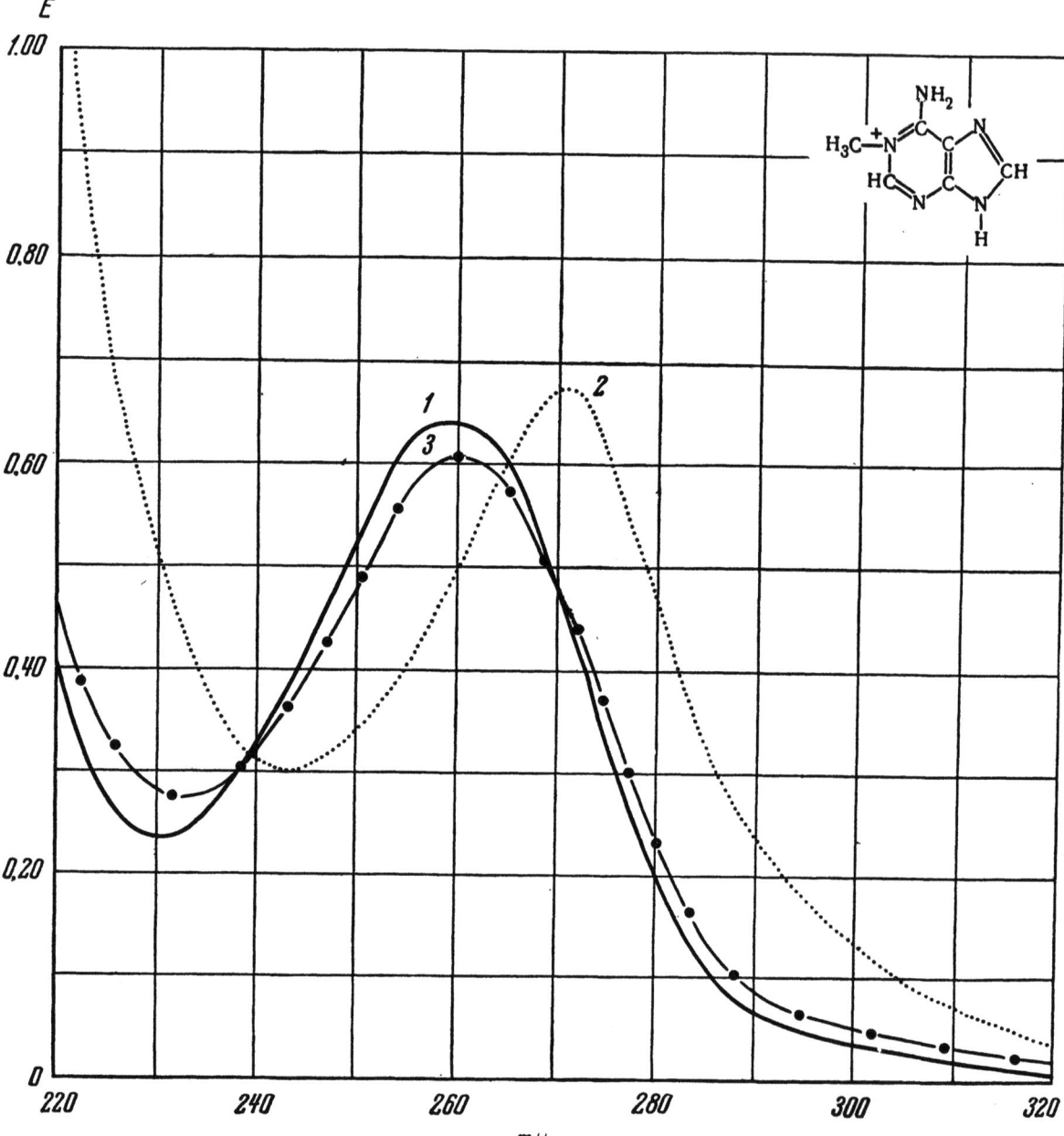

Table 5

1-METHYLADENINE

1-Methyl-6-aminopurine

$C_6H_8N_5$

Mol. wt. 150

	0.1 N HCl	0.1 N KOH
λ_{max} (mμ)	259	270
$\varepsilon_{max} \times 10^{-3}$	11.7	14.4
λ_{min} (mμ)	230	242
$\varepsilon_{min} \times 10^{-3}$	—	—
E_{250}/E_{260}	0.82	0.70
E_{270}/E_{260}	0.78	1.25
E_{280}/E_{260}	0.36	0.92
E_{290}/E_{260}	0.14	0.56

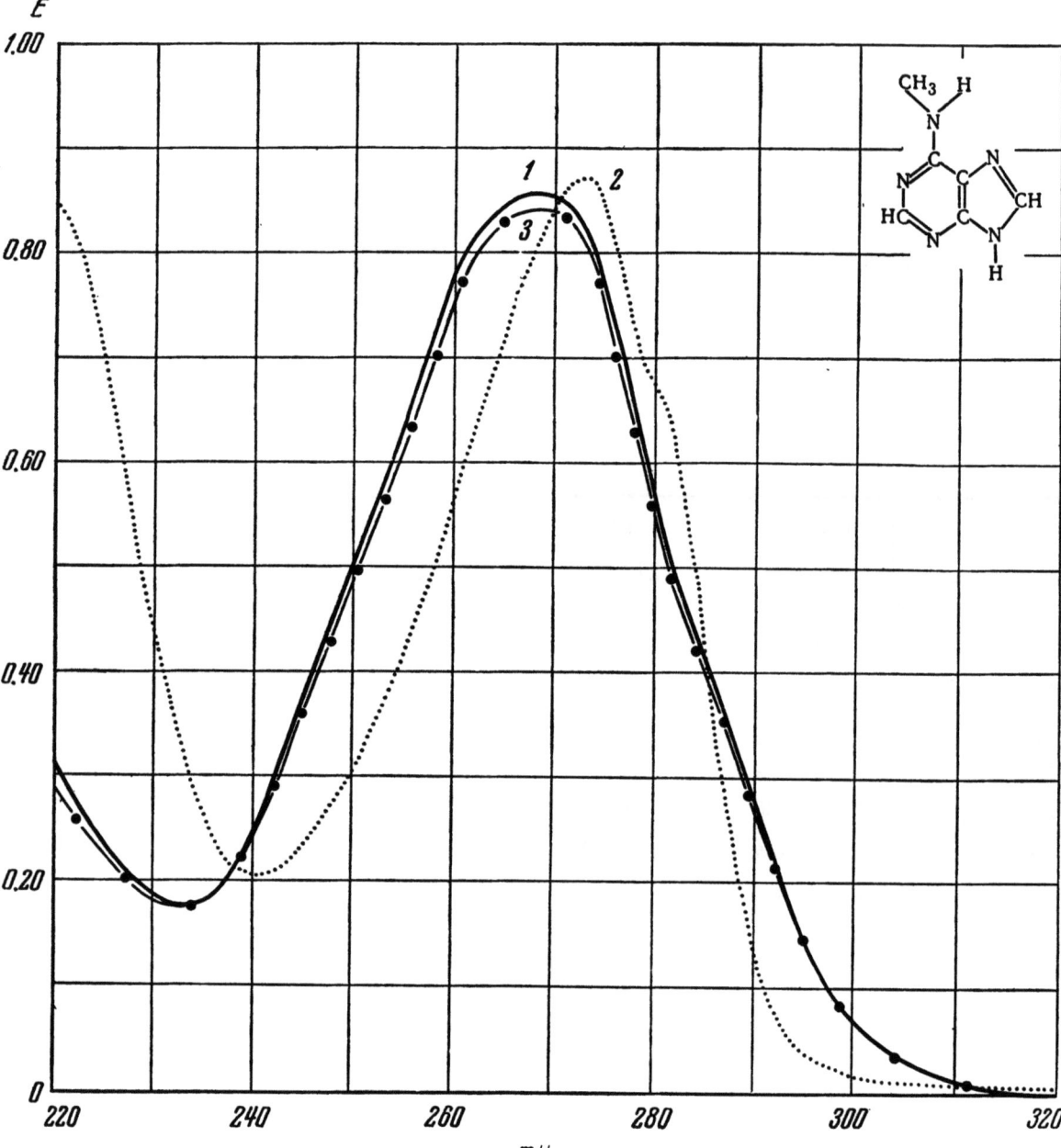

Table 6

N$_6$-METHYLADENINE

6-Methylaminopurine

C$_6$H$_7$N$_5$

Mol. wt. 149

	0.1 N HCl	0.1 N KOH
λ_{max} (mμ)	267	273
$\varepsilon_{max} \times 10^{-3}$	15.1 [12,8]	—
λ_{min} (mμ)	232	239
$\varepsilon_{min} \times 10^{-3}$	—	—
E_{250}/E_{260}	0.63	0.54
E_{270}/E_{260}	1.08	1.44
E_{280}/E_{260}	0.74	1.16
E_{290}/E_{260}	0.37	0.23

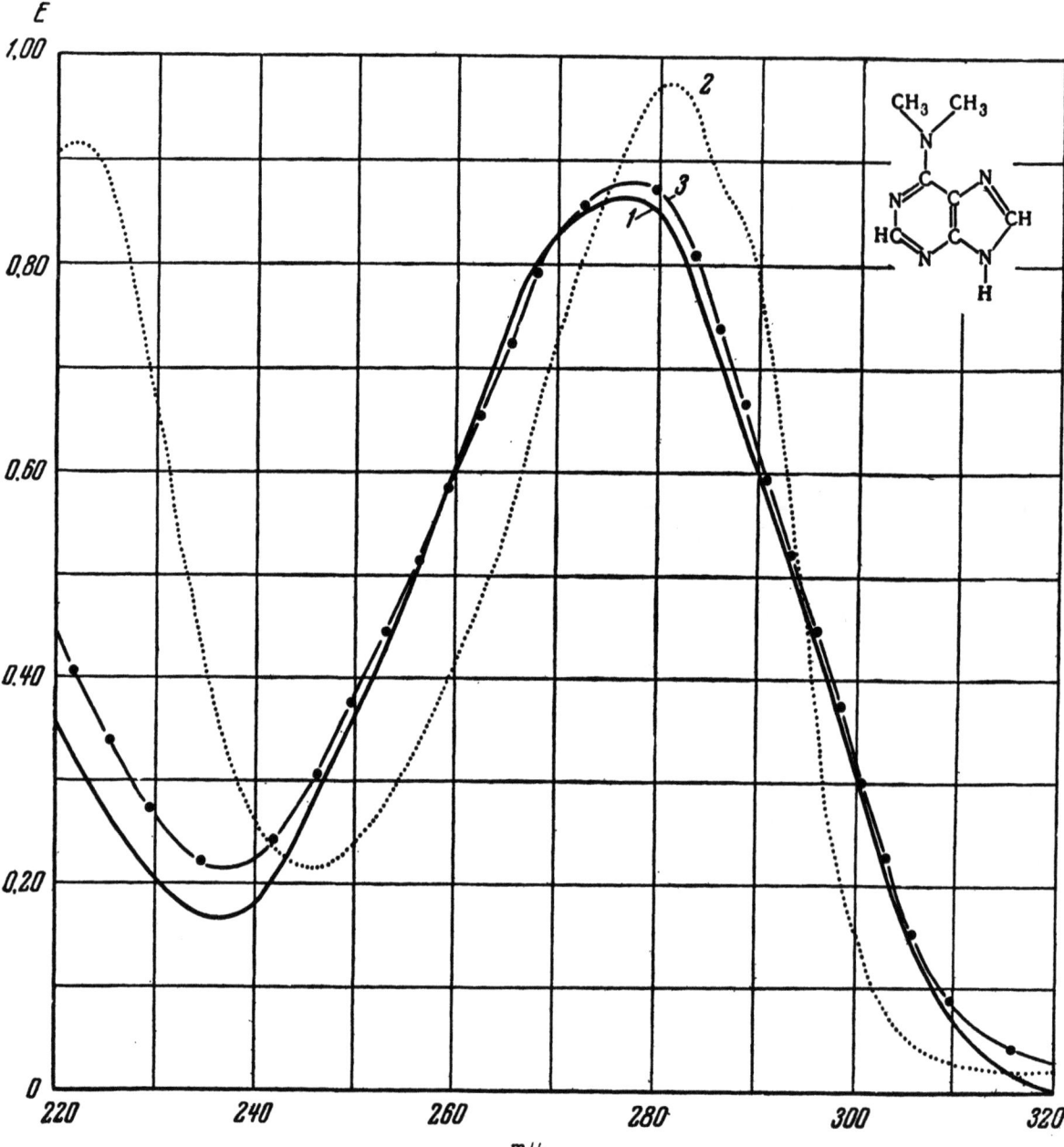

Table 7

N$_6$-DIMETHYLADENINE

6-Dimethylaminopurine

C$_7$H$_9$N$_5$

Mol. wt. 163

	0.1 N HCl	0.1 N KOH
λ_{max} (mμ)	277	281
$\varepsilon_{max} \times 10^{-3}$	15.3 [16.3]	—
λ_{min} (mμ)	236	245
$\varepsilon_{min} \times 10^{-3}$	—	—
E_{250}/E_{260}	0.58	0.57
E_{270}/E_{260}	1.33	1.70
E_{280}/E_{260}	1.37	2.26
E_{290}/E_{260}	0.97	1.85

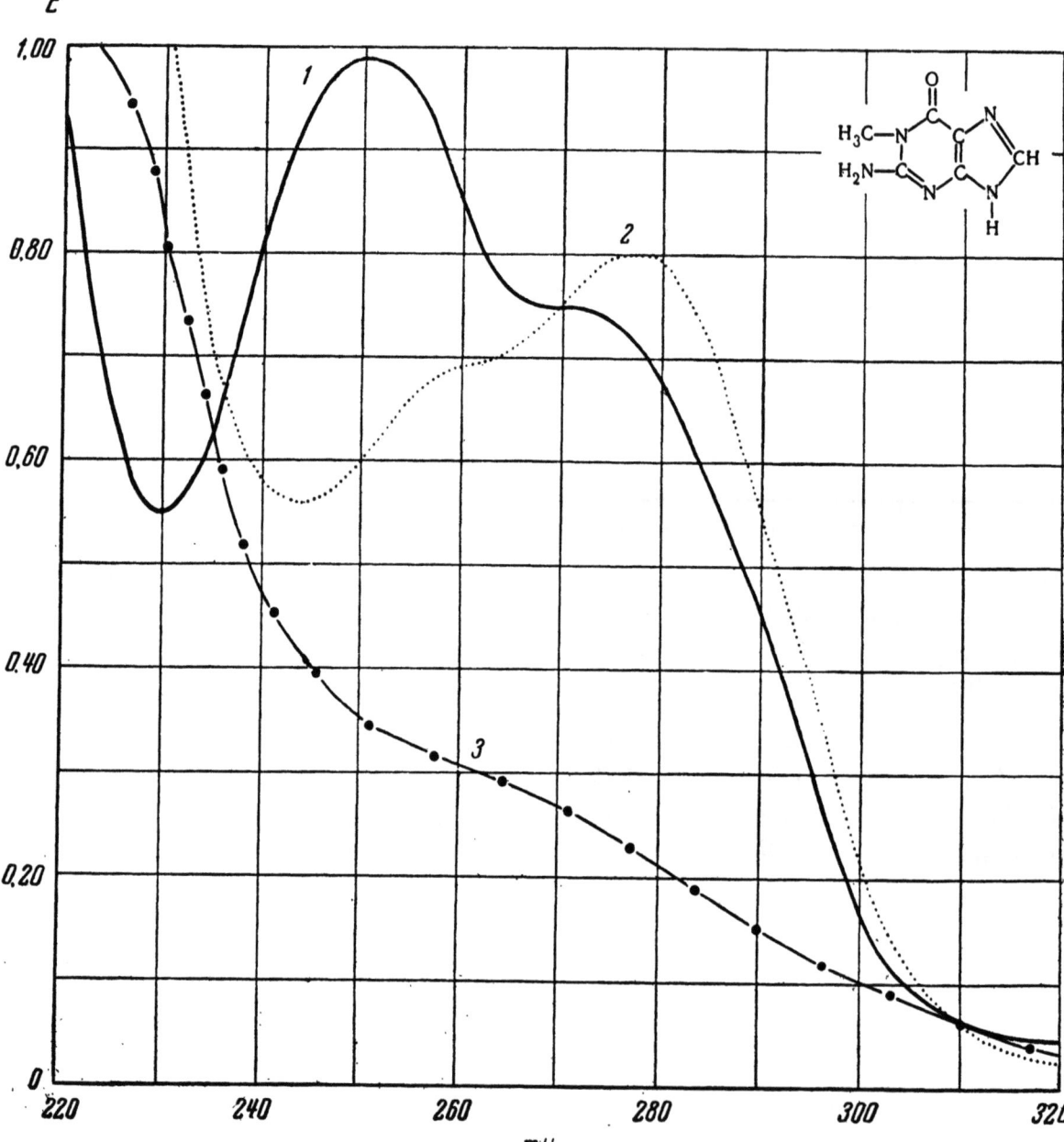

Table 8

1-METHYLGUANINE

1-Methyl-2-amino-6-hydroxypurine

$C_6H_7N_5O$

Mol. wt. 165

	0.1 N HCl	0.1 N KOH
λ_{max} (mμ)	251	277
$\varepsilon_{max} \times 10^{-3}$	—	—
λ_{min} (mμ)	228	245
$\varepsilon_{min} \times 10^{-3}$	—	—
E_{250}/E_{260}	1.16	0.87
E_{270}/E_{260}	0.89	1.09
E_{280}/E_{260}	0.80	1.14
E_{290}/E_{260}	0.55	0.80

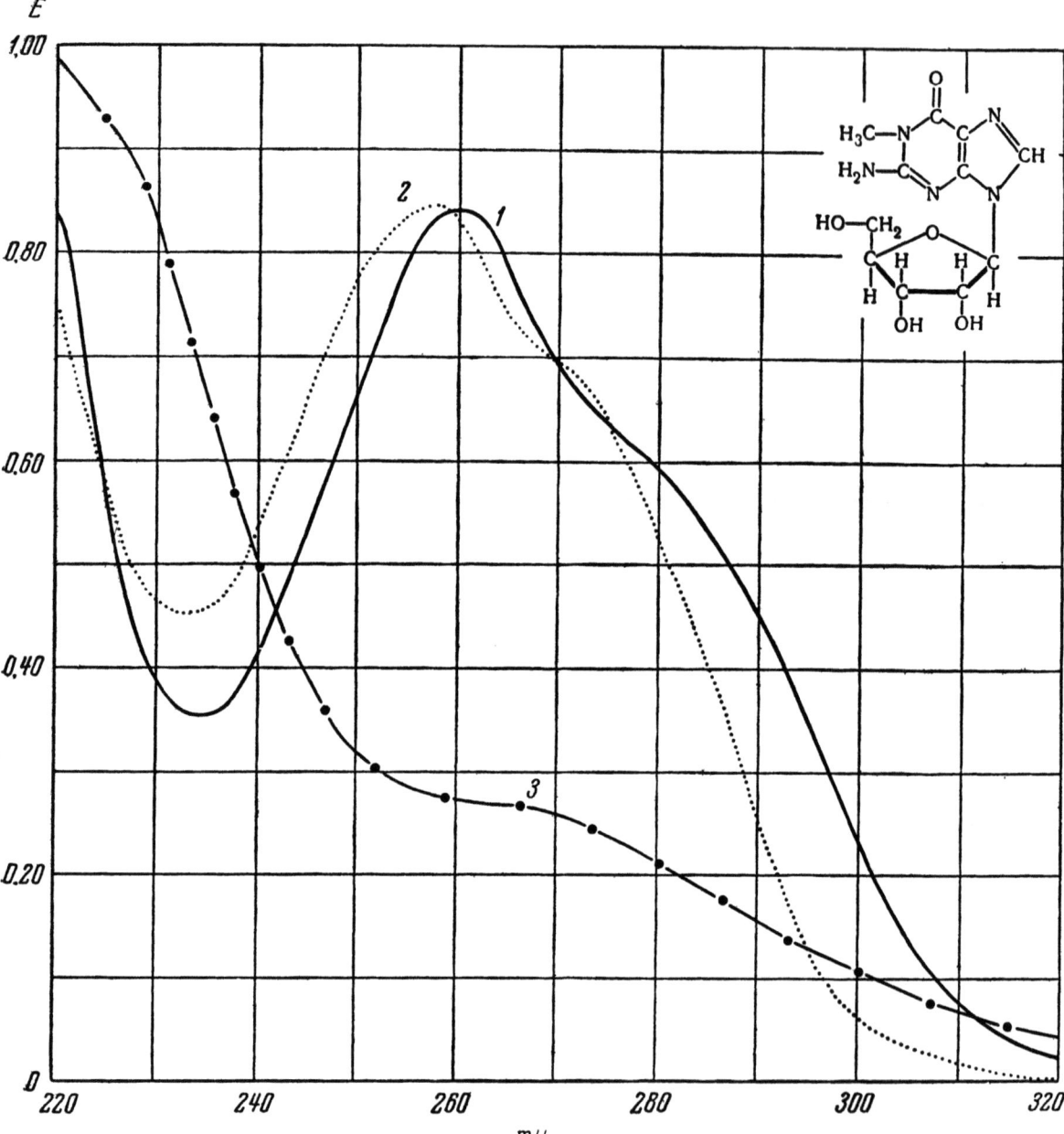

Table 9

1-METHYLGUANOSINE

1-Methyl-2-amino-
6-hydroxypurineriboside

$C_{11}H_{15}N_5O_5$

Mol. wt. 297

	0.1 N HCl	0.1 N KOH
λ_{max} (mμ)	258	256
$\varepsilon_{max} \times 10^{-3}$	12.2	—
λ_{min} (mμ)	234	232
$\varepsilon_{min} \times 10^{-3}$	—	—
E_{250}/E_{260}	0.80	0.96
E_{270}/E_{260}	0.82	0.85
E_{280}/E_{260}	0.70	0.64
E_{290}/E_{260}	0.55	0.29

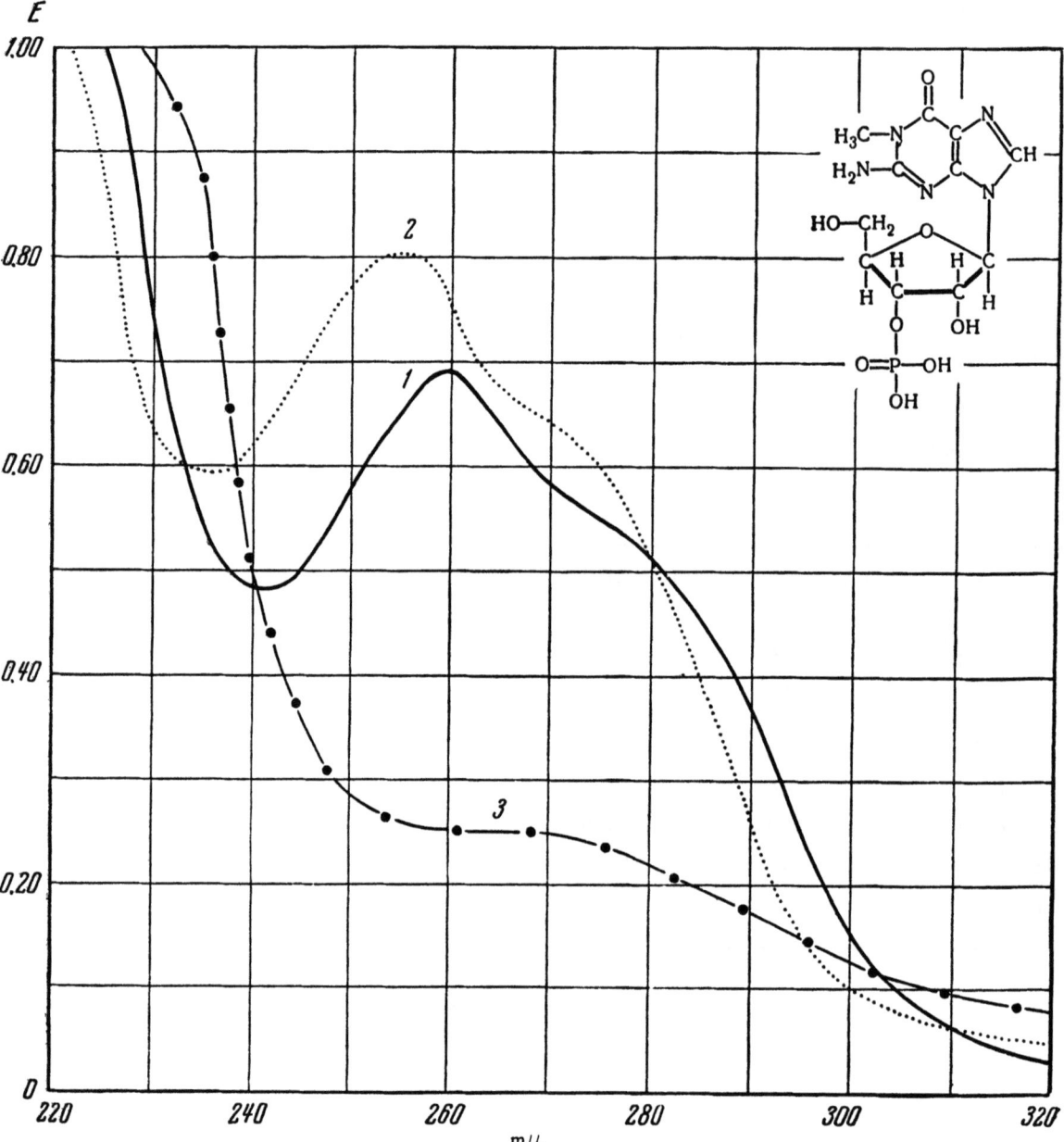

Table 10

1-METHYLGUANOSINE-3'-PHOSPHATE
1-Methyl-2-amino-
6-hydroxypurineriboside-3'-phosphate
$C_{11}H_{16}N_5O_8P$
Mol. wt. 377

	0.1 N HCl	0.1 N KOH
λ_{max} (mμ)	259	256
$\varepsilon_{max} \times 10^{-3}$	—	—
λ_{min} (mμ)	240	235
$\varepsilon_{min} \times 10^{-3}$	—	—
E_{250}/E_{260}	0.86	1.03
E_{270}/E_{260}	0.82	0.85
E_{280}/E_{260}	0.74	0.68
E_{290}/E_{260}	0.58	0.35

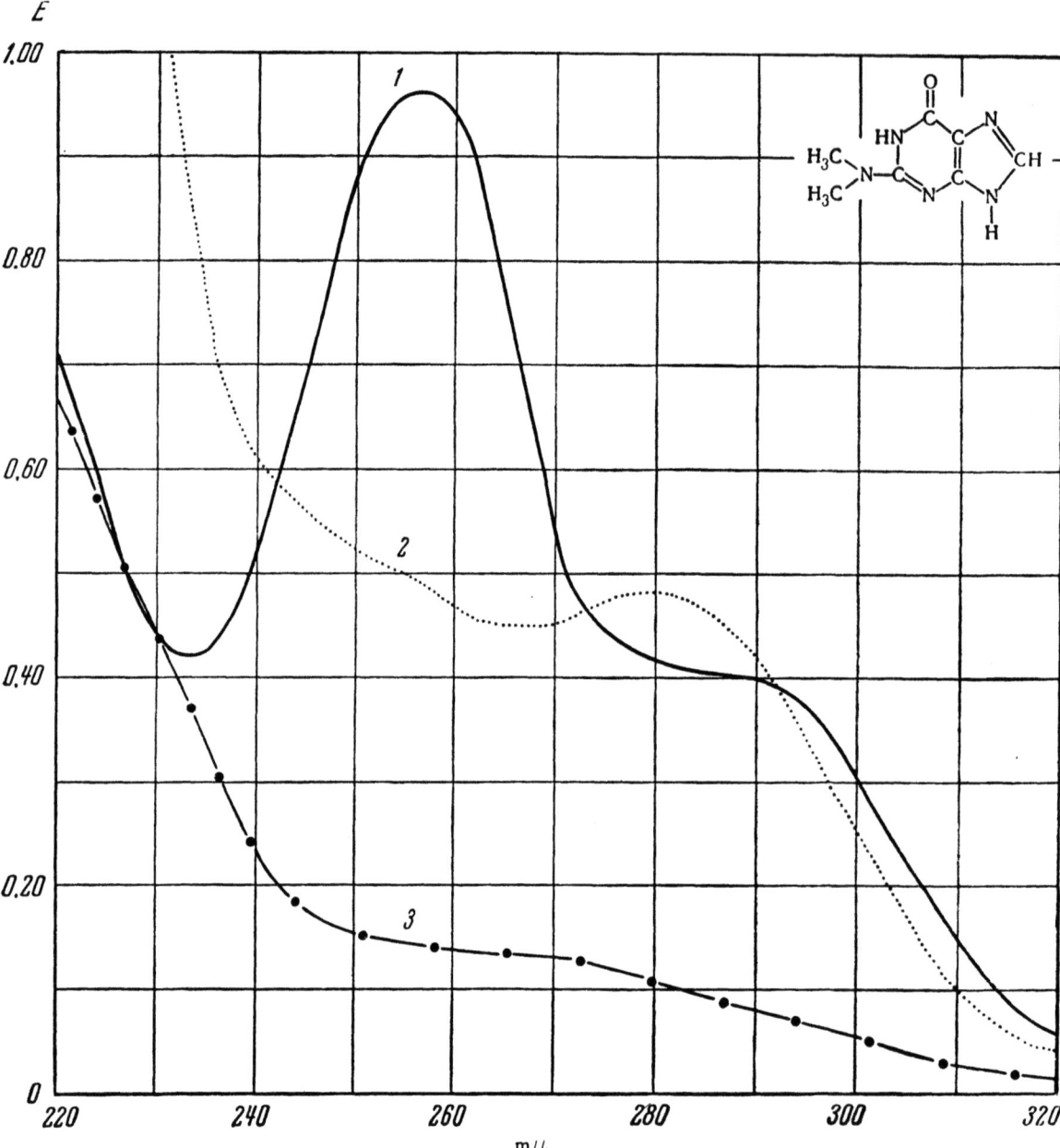

Table 11

N$_2$-DIMETHYLGUANINE
2-Dimethylamino-6-hydroxypurine
C$_7$H$_9$N$_5$O
Mol. wt. 179

	0.1 N HCl	0.1 N KOH
λ_{max} (mμ)	256	277
$\varepsilon_{max} \times 10^{-3}$	14.5	8.0
λ_{min} (mμ)	235	265
$\varepsilon_{min} \times 10^{-3}$	6.9	7.1
E_{250}/E_{260}	0.92	1.1
E_{270}/E_{260}	0.56	0.96
E_{280}/E_{260}	0.44	1.0
E_{290}/E_{260}	0.42	0.90

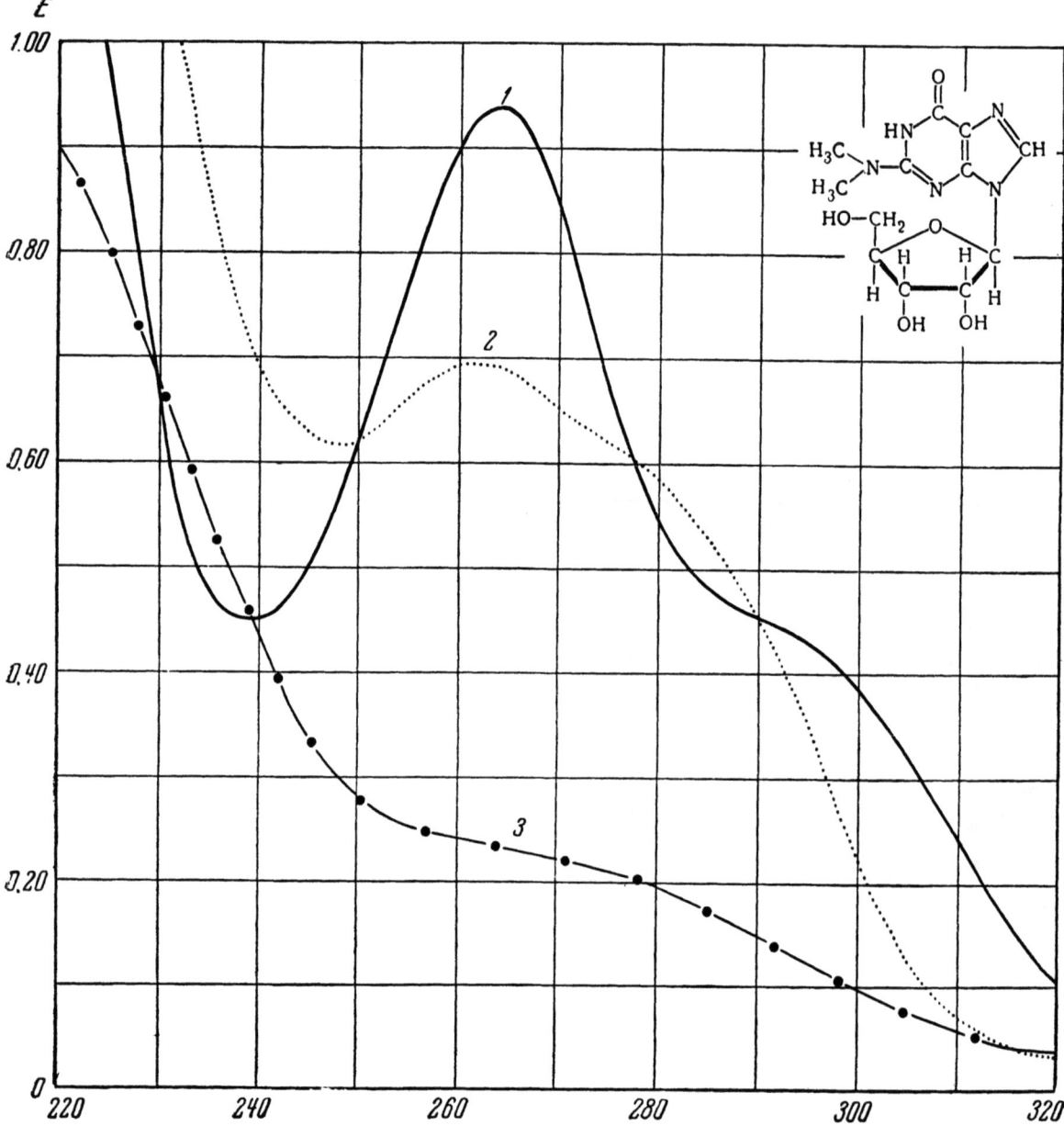

Table 12

N₂-DIMETHYLGUANOSINE

2-Dimethylamino-
6-hydroxypurineriboside

$C_{12}H_{17}N_5O_5$

Mol. wt. 311

	0.1 N HCl	0.1 N KOH
λ_{max} (mμ)	265	262
$\varepsilon_{max} \times 10^{-3}$	15.1	—
λ_{min} (mμ)	240	248
$\varepsilon_{min} \times 10^{-3}$	—	—
E_{250}/E_{260}	0.70	0.90
E_{270}/E_{260}	0.95	0.94
E_{280}/E_{260}	0.60	0.84
E_{290}/E_{260}	0.50	0.67

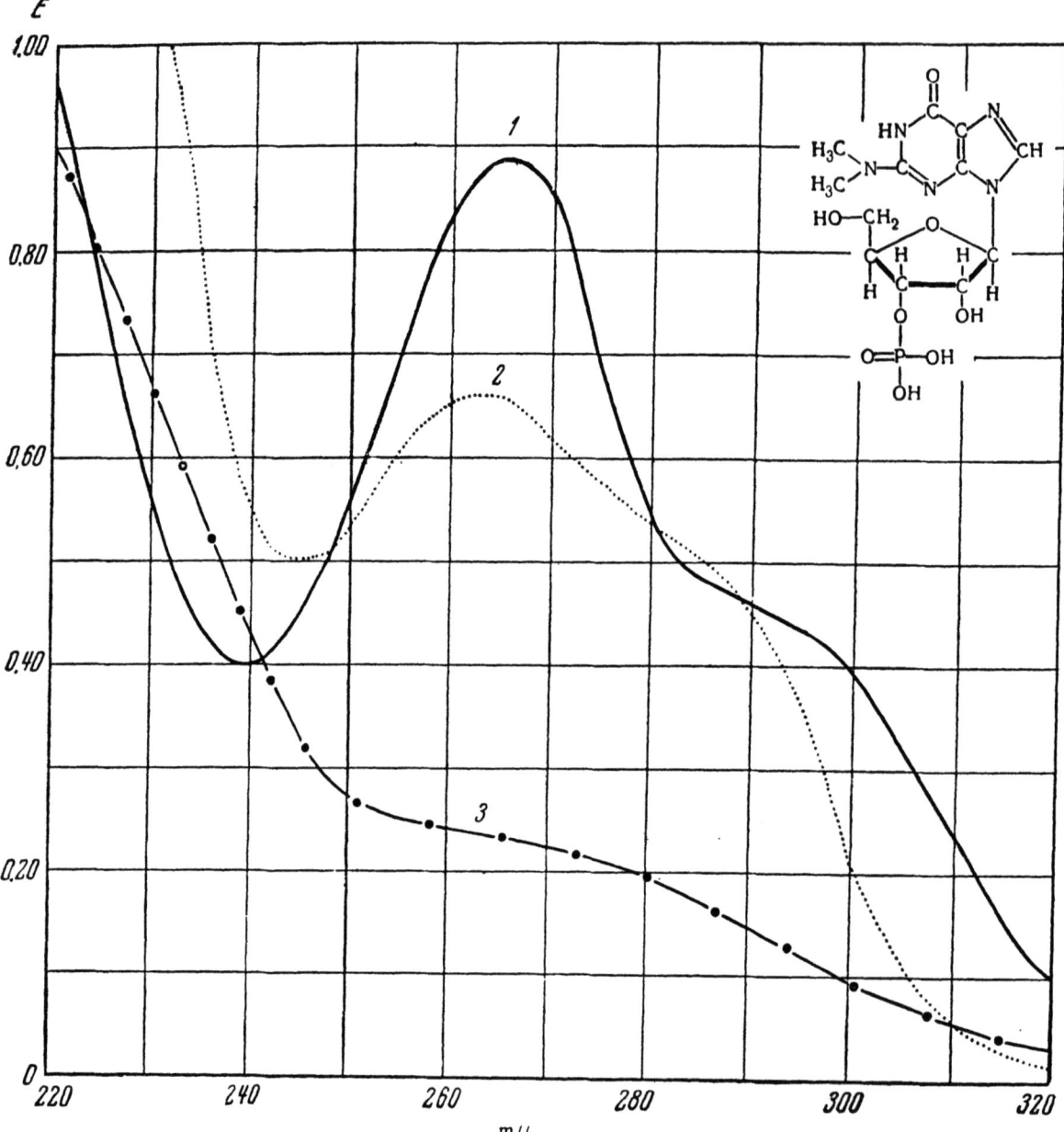

Table 13

N$_2$-DIMETHYLGUANOSINE-3'-PHOSPHATE
2-Dimethylamino-6-hydroxypurineriboside-
3'-phosphate
C$_{12}$H$_{18}$N$_5$O$_8$P
Mol. wt. 391

	0.1 N HCl	0.1 N KOH
λ_{max} (mμ)	265	263
$\varepsilon_{max} \times 10^{-3}$	—	—
λ_{min} (mμ)	240	244
$\varepsilon_{min} \times 10^{-3}$	—	—
E_{250}/E_{260}	0.67	0.82
E_{270}/E_{260}	1.0	0.94
E_{280}/E_{260}	0.65	0.82
E_{290}/E_{260}	0.54	0.67

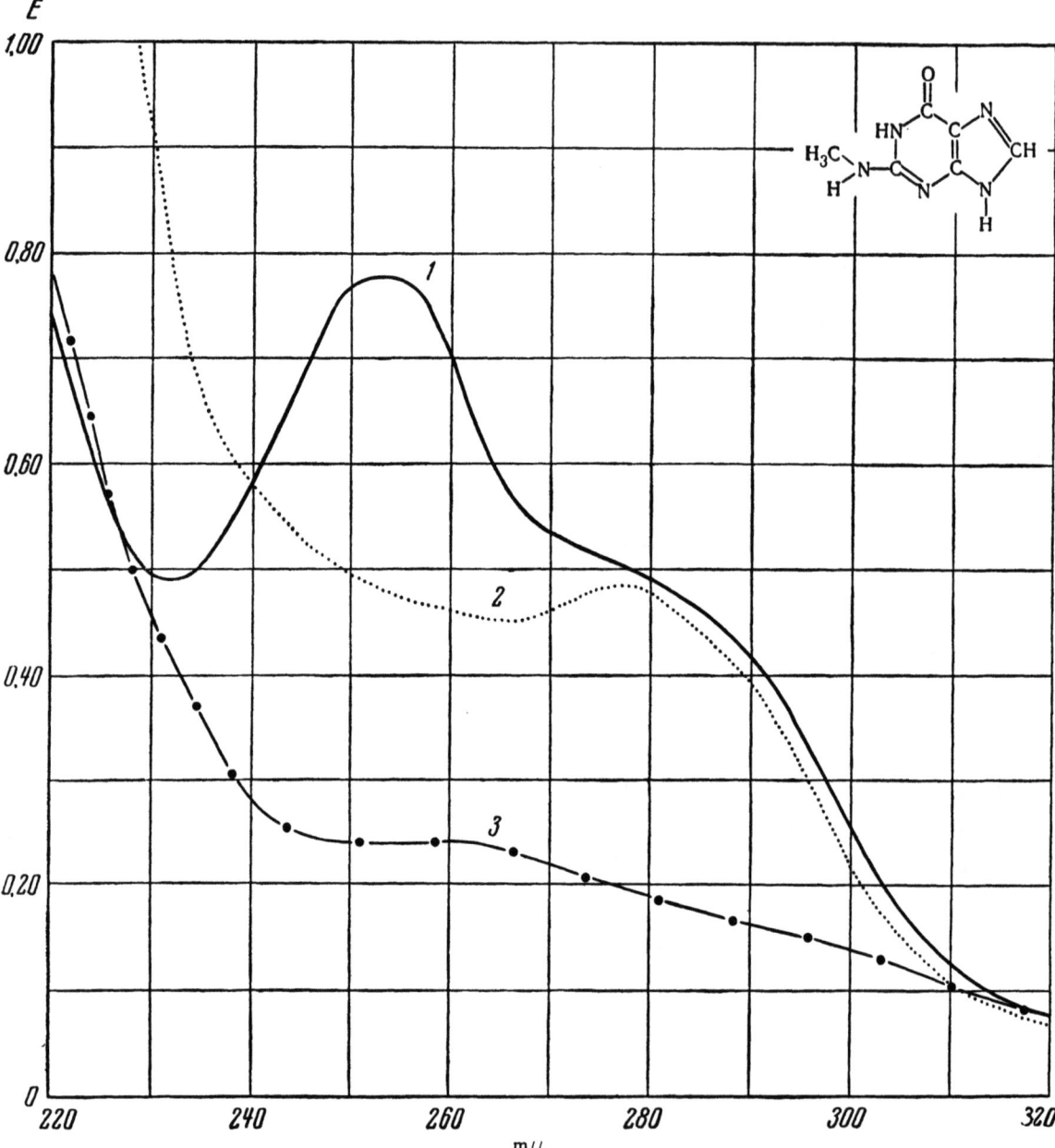

Table 14

N$_2$-METHYLGUANINE

2-Methylamino-6-hydroxypurine

C$_6$H$_7$N$_5$O

Mol. wt. 165

	0.1 N HCl	0.1 N KOH
λ_{max} (mμ)	252	278
$\varepsilon_{max} \times 10^{-3}$	12.3	7.2
λ_{min} (mμ)	232	263
$\varepsilon_{min} \times 10^{-3}$	—	6.1
E_{250}/E_{260}	1.1	1.06
E_{270}/E_{260}	0.76	1.0
E_{280}/E_{260}	0.70	1.04
E_{290}/E_{260}	0.60	0.85

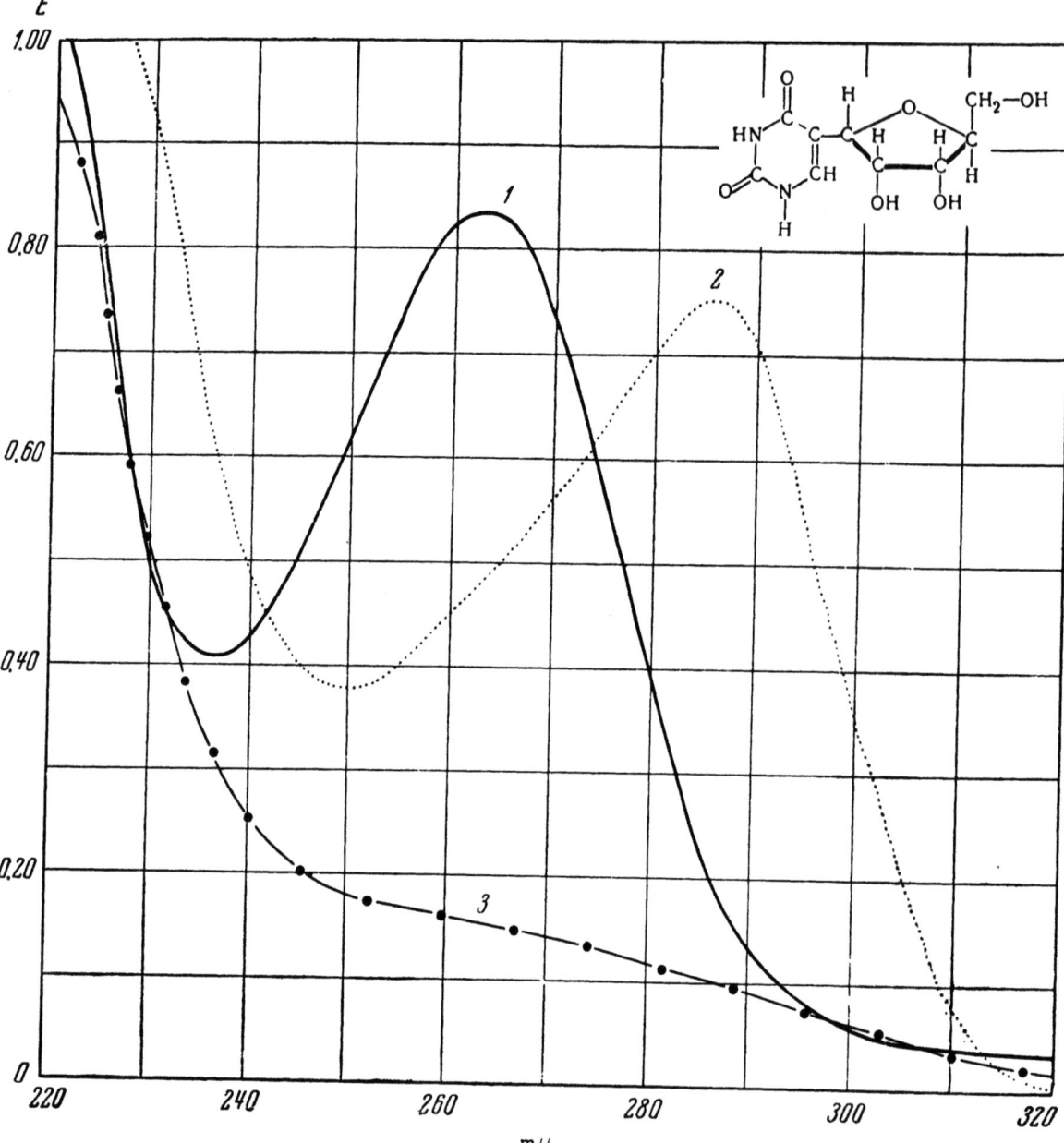

Table 15

PSEUDOURIDINE [ψ URIDINE]

5-Ribosyluracil

$C_9H_{12}N_2O_6$

Mol. wt. 244

	0.1 N HCl	0.1 N KOH
λ_{max} (mμ)	263	286
$\varepsilon_{max} \times 10^{-3}$	7.5	7.3
λ_{min} (mμ)	233	245
$\varepsilon_{min} \times 10^{-3}$	2.0	1.7
E_{250}/E_{260}	0.77	0.87
E_{270}/E_{260}	0.92	1.25
E_{280}/E_{260}	0.53	1.54
E_{290}/E_{260}	0.20	1.54

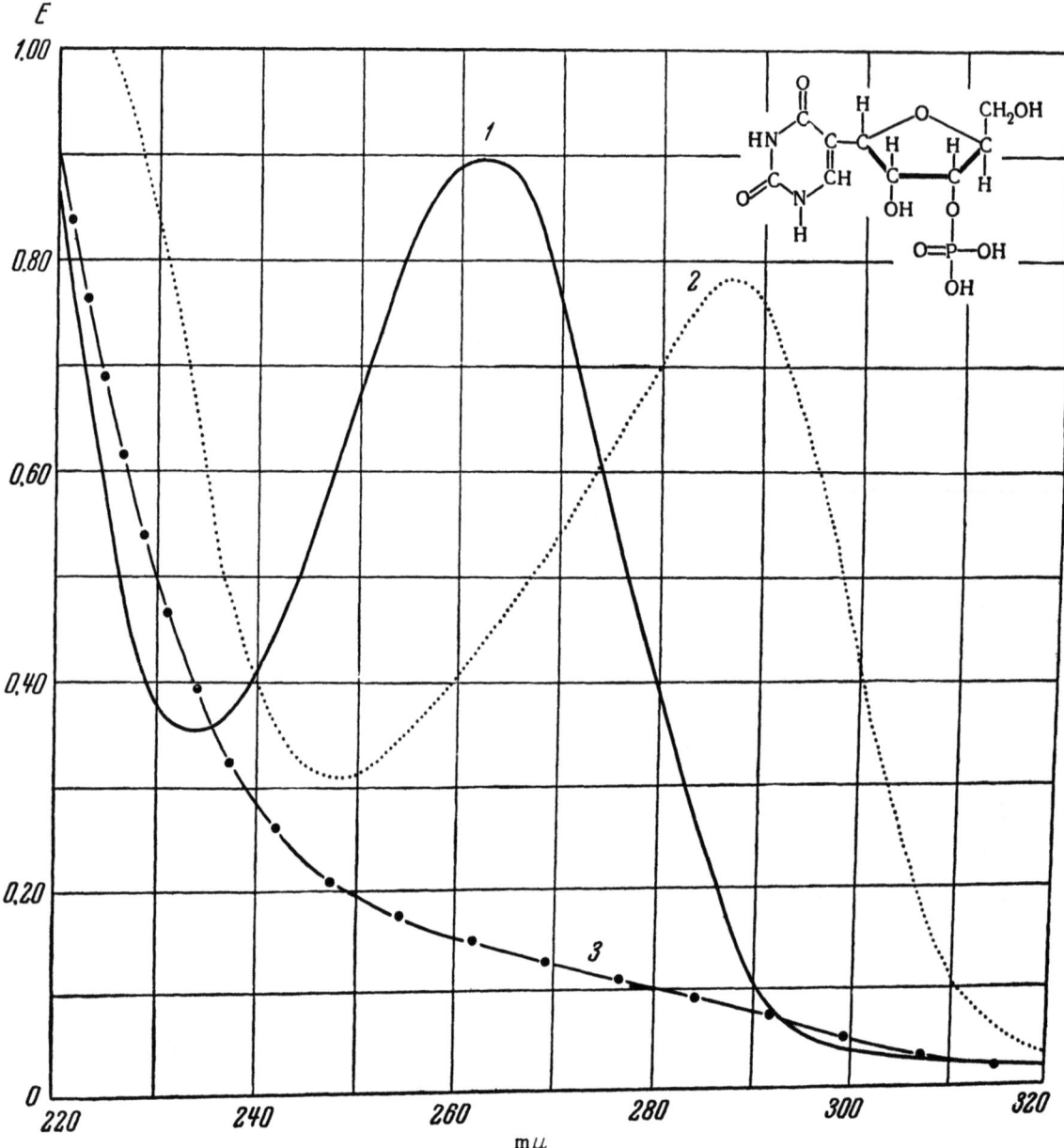

Table 16

PSEUDOURIDYLIC ACID [ψ UMP]
5-Ribosyluracil-3'-phosphate
$C_9H_{13}N_2O_9P$
Mol. wt. 324

	0.1 N HCl	0.1 N KOH
λ_{max} (mμ)	263	286
$\varepsilon_{max} \times 10^{-3}$	8.4	8.4
λ_{min} (mμ)	233	246
$\varepsilon_{min} \times 10^{-3}$	2.3	2.1
E_{250}/E_{260}	0.76	0.77
E_{270}/E_{260}	0.88	1.32
E_{280}/E_{260}	0.46	1.70
E_{290}/E_{260}	0.13	1.83

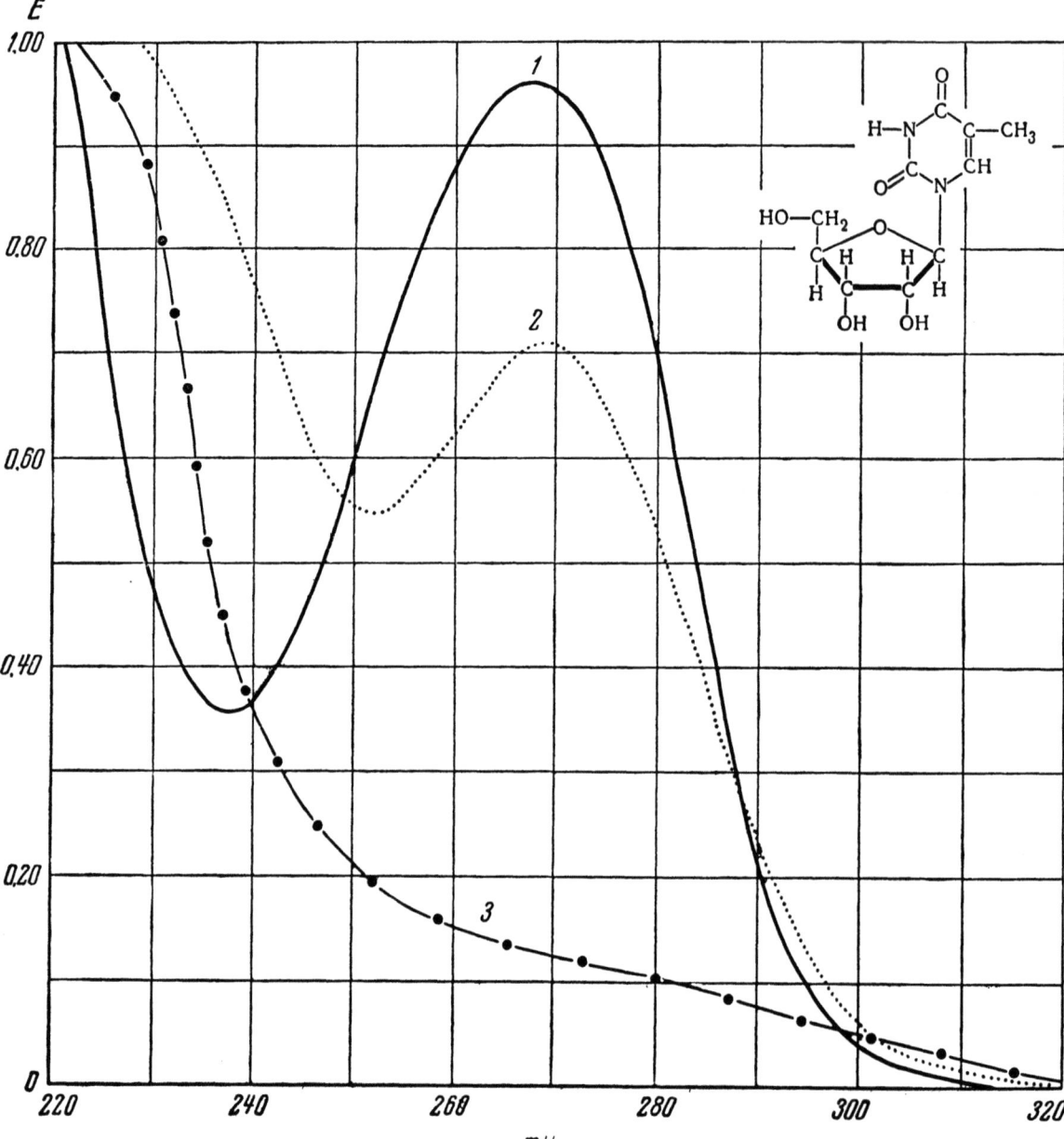

Table 17

THYMINE RIBOSIDE

5-Methyl-3-ribosyluracil

$C_{10}H_{14}N_2O_6$

Mol. wt. 258

	0.1 N HCl	0.1 N KOH
λ_{max} (mμ)	267	268
$\varepsilon_{max} \times 10^{-3}$	9.65	7.38
λ_{min} (mμ)	238	250
$\varepsilon_{min} \times 10^{-3}$	—	—
E_{250}/E_{260}	0.68	0.88
E_{270}/E_{260}	1.06	1.09
E_{280}/E_{260}	0.73	0.81
E_{290}/E_{260}	0.29	0.41

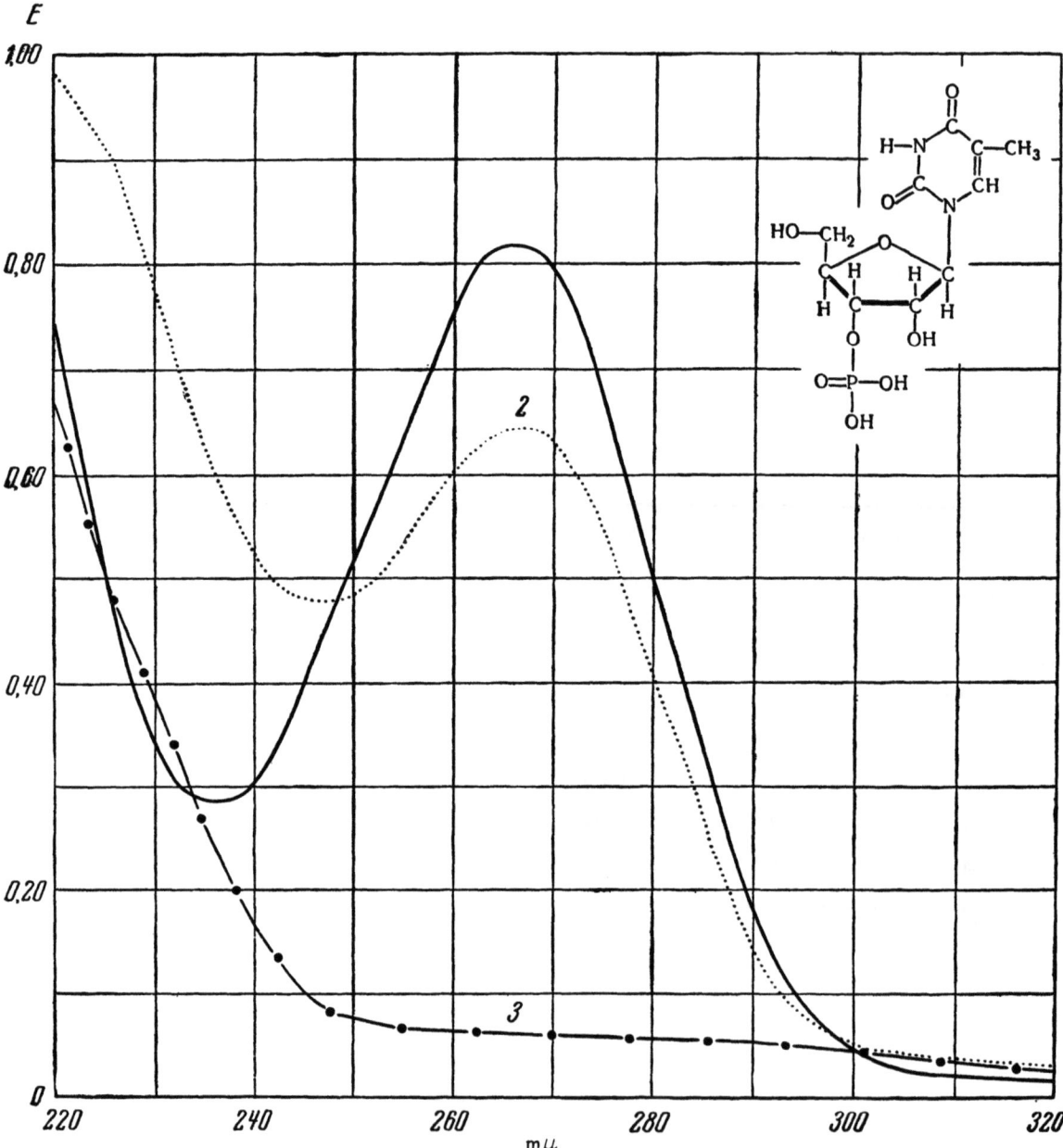

Table 18

THYMINE RIBOSIDE-3'-PHOSPHATE [TMP]
5-Methyl-3-ribosyluracil-3'-phosphate

$C_{10}H_{15}N_2O_9P$

Mol. wt. 338

	0.1 N HCl	0.1 N KOH
λ_{max} (mμ)	267	268
$\varepsilon_{max} \times 10^{-3}$	9.8	—
λ_{min} (mμ)	235	247
$\varepsilon_{min} \times 10^{-3}$	—	—
E_{250}/E_{260}	0.68	0.79
E_{270}/E_{260}	1.05	1.04
E_{280}/E_{260}	0.66	0.69
E_{290}/E_{260}	0.23	0.23

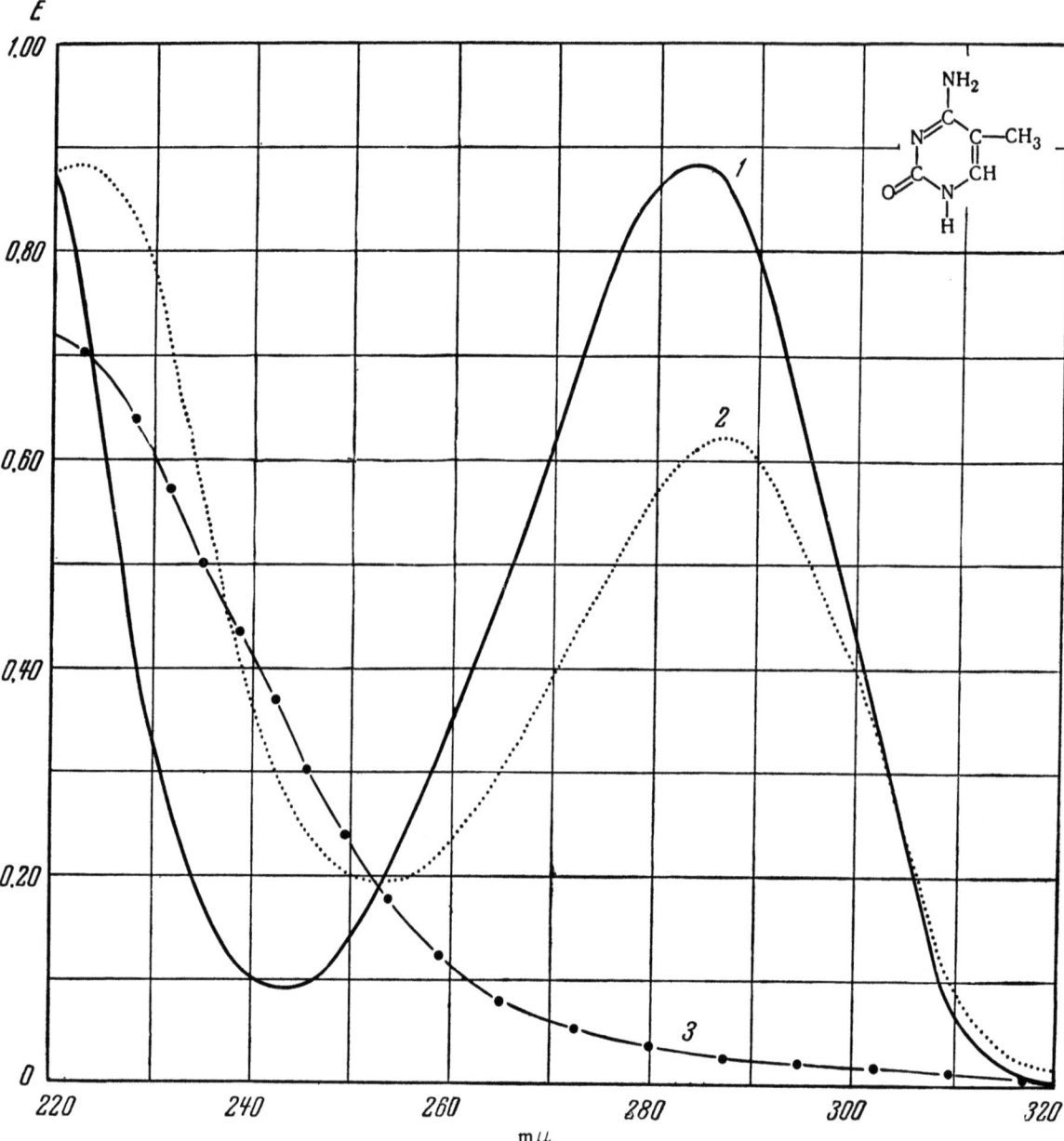

Table 19

5-METHYLCYTOSINE

2-Hydroxy-5-methyl-

6-aminopyrimidine

$C_5H_7N_3O$

Mol. wt. 125

	0.1 N HCl	0.1 N KOH
λ_{max} (mμ)	283	287
$\varepsilon_{max} \times 10^{-3}$	9.81 [8.52]	6.87
λ_{min} (mμ)	242	253
$\varepsilon_{min} \times 10^{-3}$	0.95	1.67
E_{250}/E_{260}	0.42	0.87
E_{270}/E_{260}	1.76	1.68
E_{280}/E_{260}	2.47	2.46
E_{290}/E_{260}	2.26	2.59

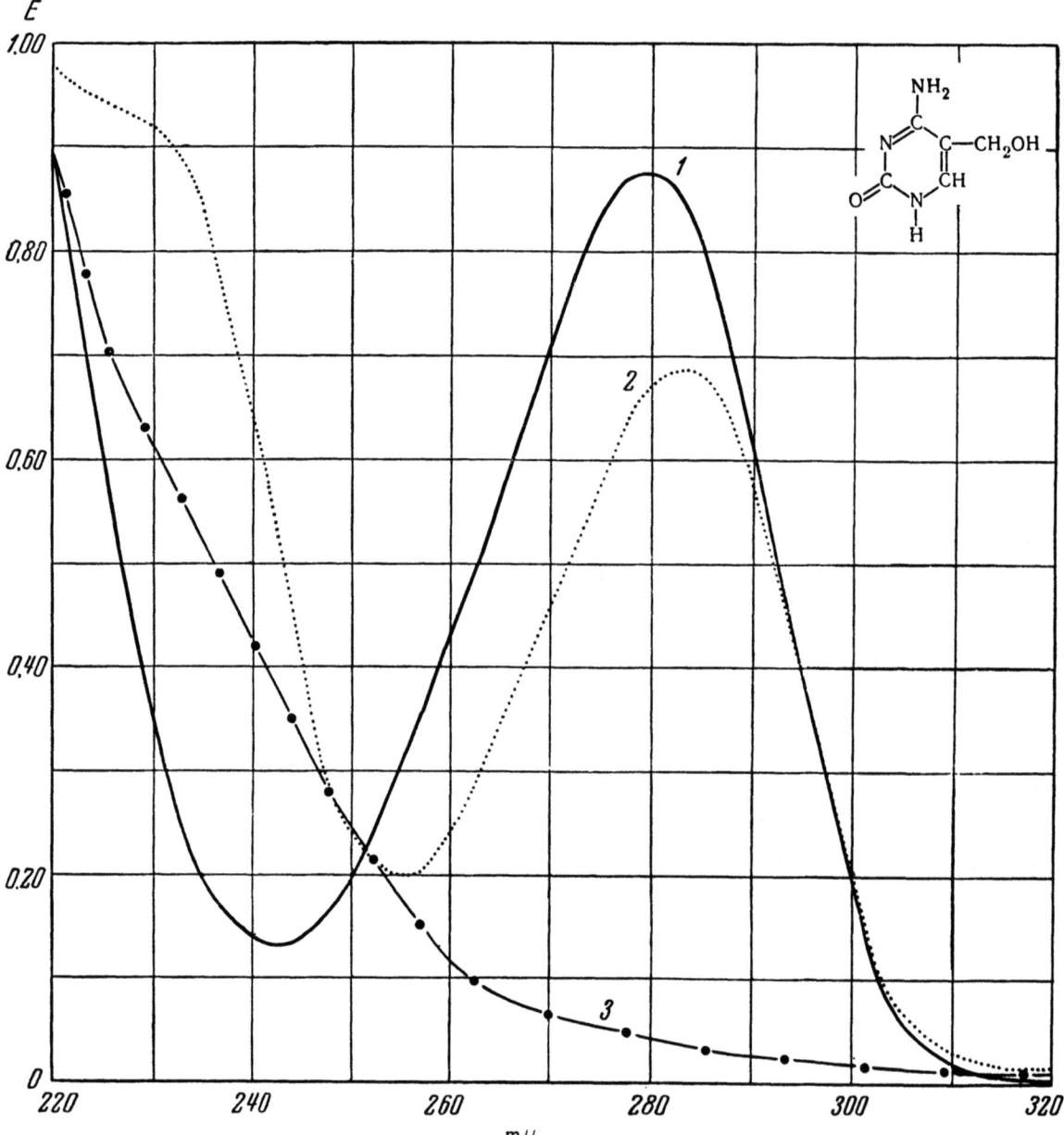

Table 20

5-HYDROXYMETHYLCYTOSINE
2-Hydroxy-5-hydroxymethyl-
6-aminopyrimidine
$C_5H_7N_3O_2$
Mol. wt. 141

	0.1 N HCl	0.1 N KOH
λ_{max} (mμ)	279	283
$\varepsilon_{max} \times 10^{-3}$	9.7 [9.24]	7.59
λ_{min} (mμ)	241	254
$\varepsilon_{min} \times 10^{-3}$	1.23	1.89
E_{250}/E_{260}	0.45	1.00
E_{270}/E_{260}	1.65	1.80
E_{280}/E_{260}	2.00	2.72
E_{290}/E_{260}	1.40	2.40

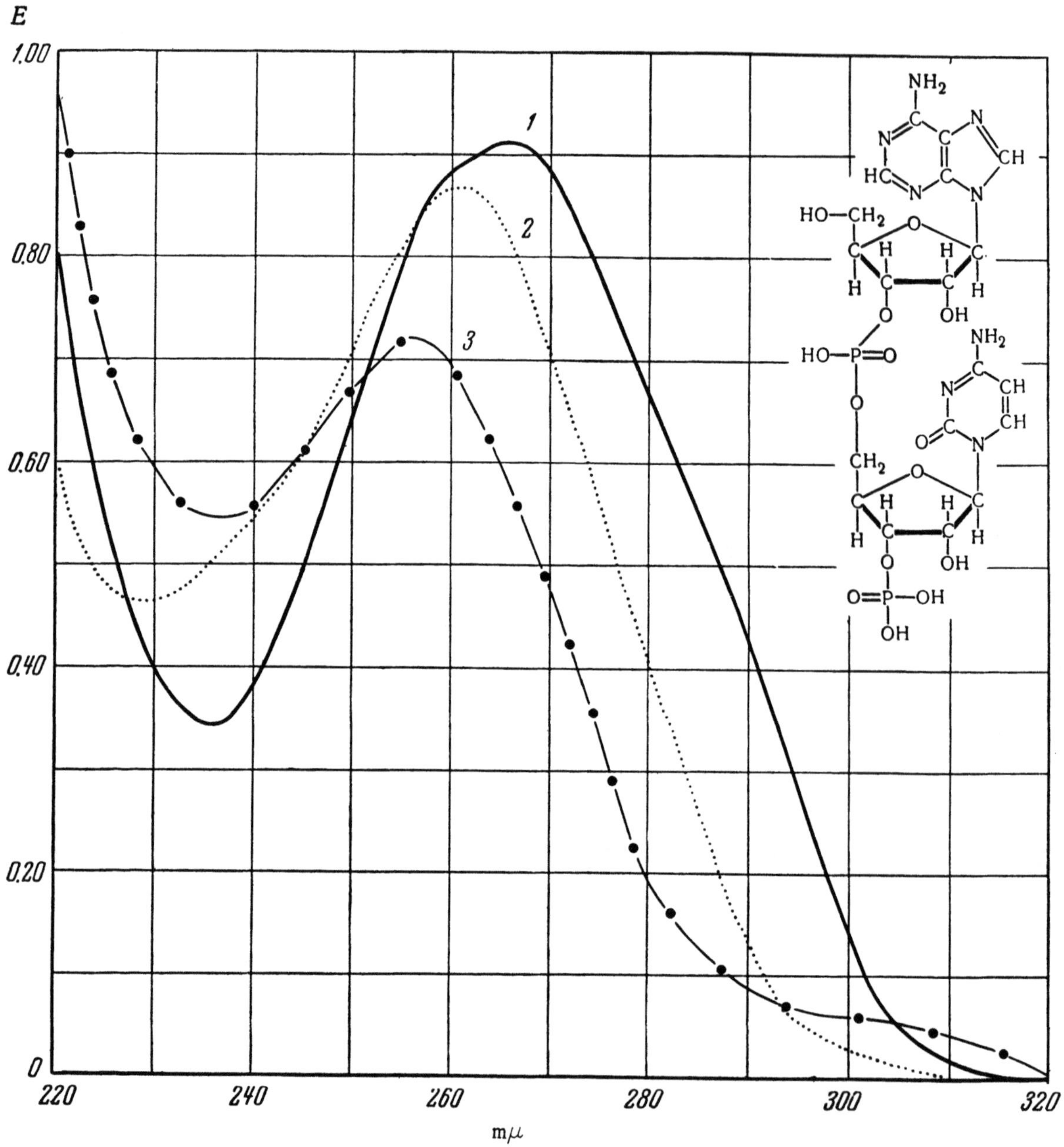

E
1.00
0.80
0.60
0.40
0.20
0
220
240
260
280
300
320
mμ
1
2
3
NH2
N
C
CH
HC
HO—CH2
O
H
OH
HO—P=O
CH2
O=P—OH

Table 21

ADENYLYL-3',5'-CYTIDINE-3'-PHOSPHATE [ApCp]

$C_{19}H_{26}N_6O_{14}P_2$

Mol. wt. 652

	0.1 N HCl	0.1 N KOH
λ_{max} (mμ)	267	262
$\varepsilon_{260} \times 10^{-3}$	21.2	—
λ_{min} (mμ)	236	230
$\varepsilon_{min} \times 10^{-3}$	—	—
E_{250}/E_{260}	0.73	0.80
E_{270}/E_{260}	1.00	0.80
E_{280}/E_{260}	0.80	0.44
E_{290}/E_{260}	0.50	0.14

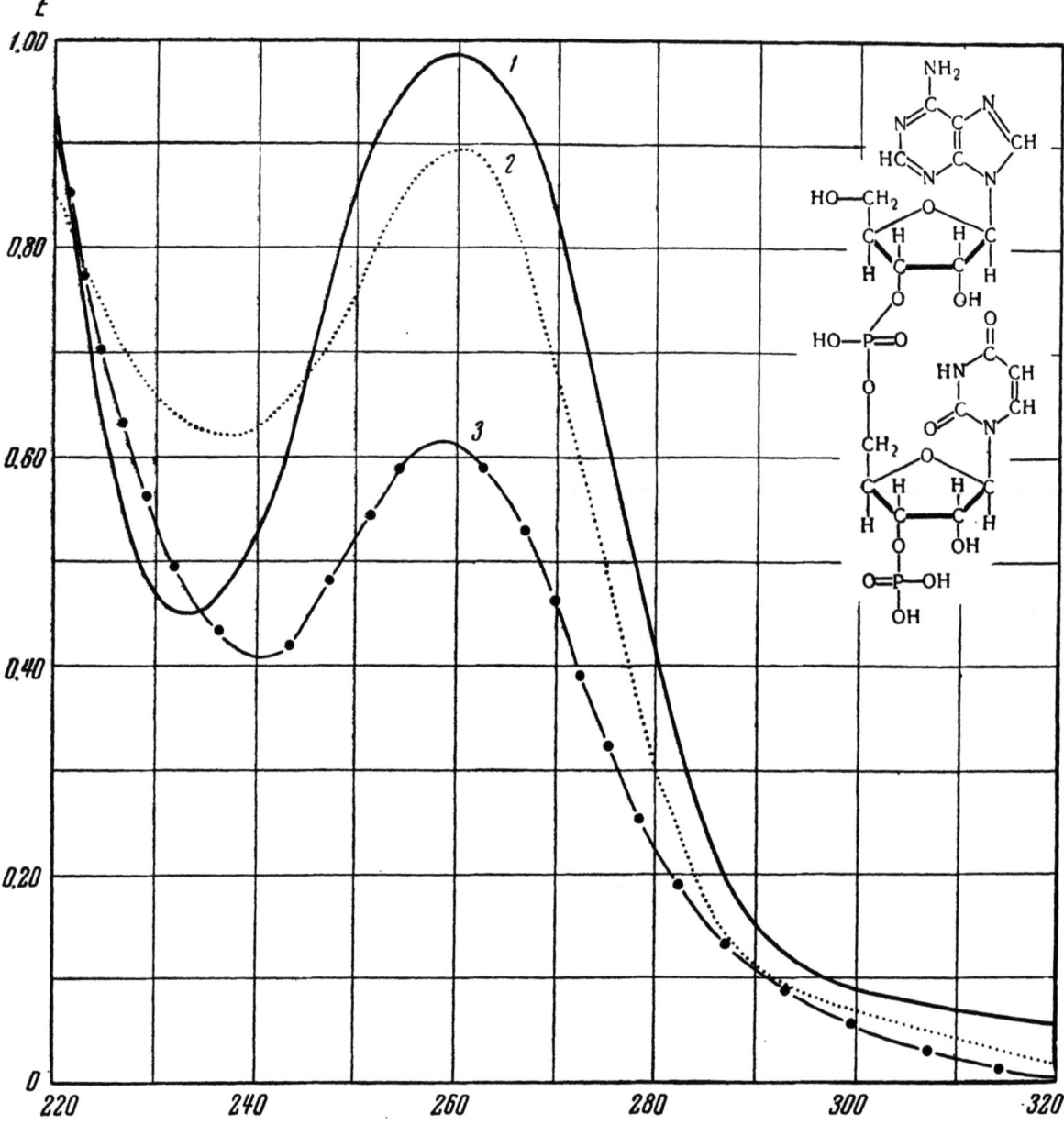

Table 22

ADENYLYL-3', 5'-URIDINE-3'-PHOSPHATE [ApUp]

$C_{19}H_{25}N_7O_{15}P_2$

Mol. wt. 653

	0.1 N HCl	0.1 N KOH
λ_{max} (mμ)	260	262
$\varepsilon_{260} \times 10^{-3}$	—	—
λ_{min} (mμ)	233	236
$\varepsilon_{min} \times 10^{-3}$	—	—
E_{250}/E_{260}	0.88	0.85
E_{270}/E_{260}	0.84	0.77
E_{280}/E_{260}	0.40	0.30
E_{290}/E_{260}	0.15	0.10

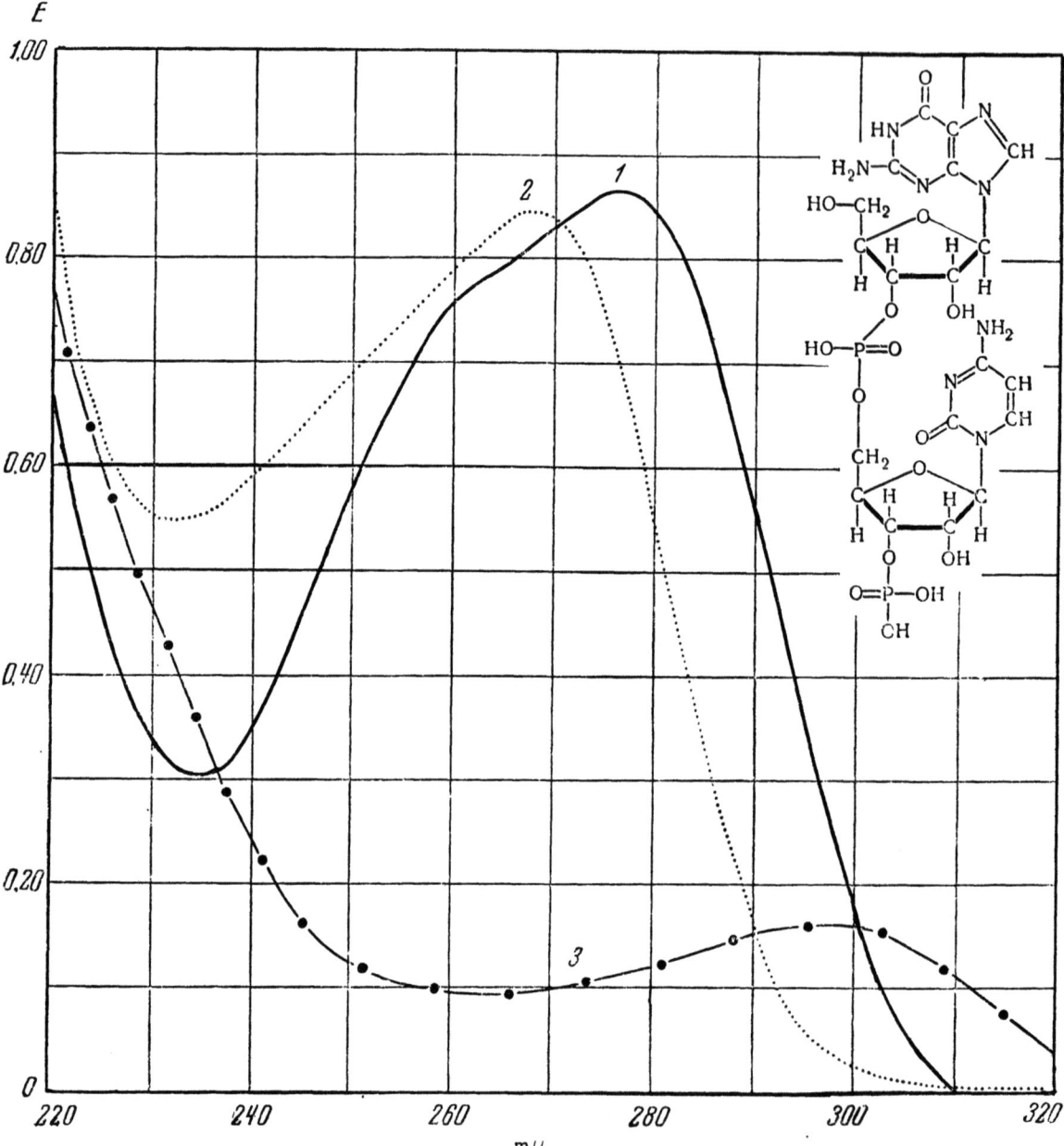

Table 23

GUANYLYL-3',5'-CYTIDINE-3'-PHOSPHATE [GpCp]

$C_{19}H_{26}N_8O_{15}P_2$

Mol. wt. 668

	0.1 N HCl	0.1 N KOH
λ_{max} (mμ)	278	269
$\varepsilon_{260} \times 10^{-3}$	19.2	—
λ_{min} (mμ)	235	231
$\varepsilon_{min} \times 10^{-3}$	—	—
E_{250}/E_{260}	0.76	0.86
E_{270}/E_{260}	1.09	1.05
E_{280}/E_{260}	1.12	0.71
E_{290}/E_{260}	0.74	0.21

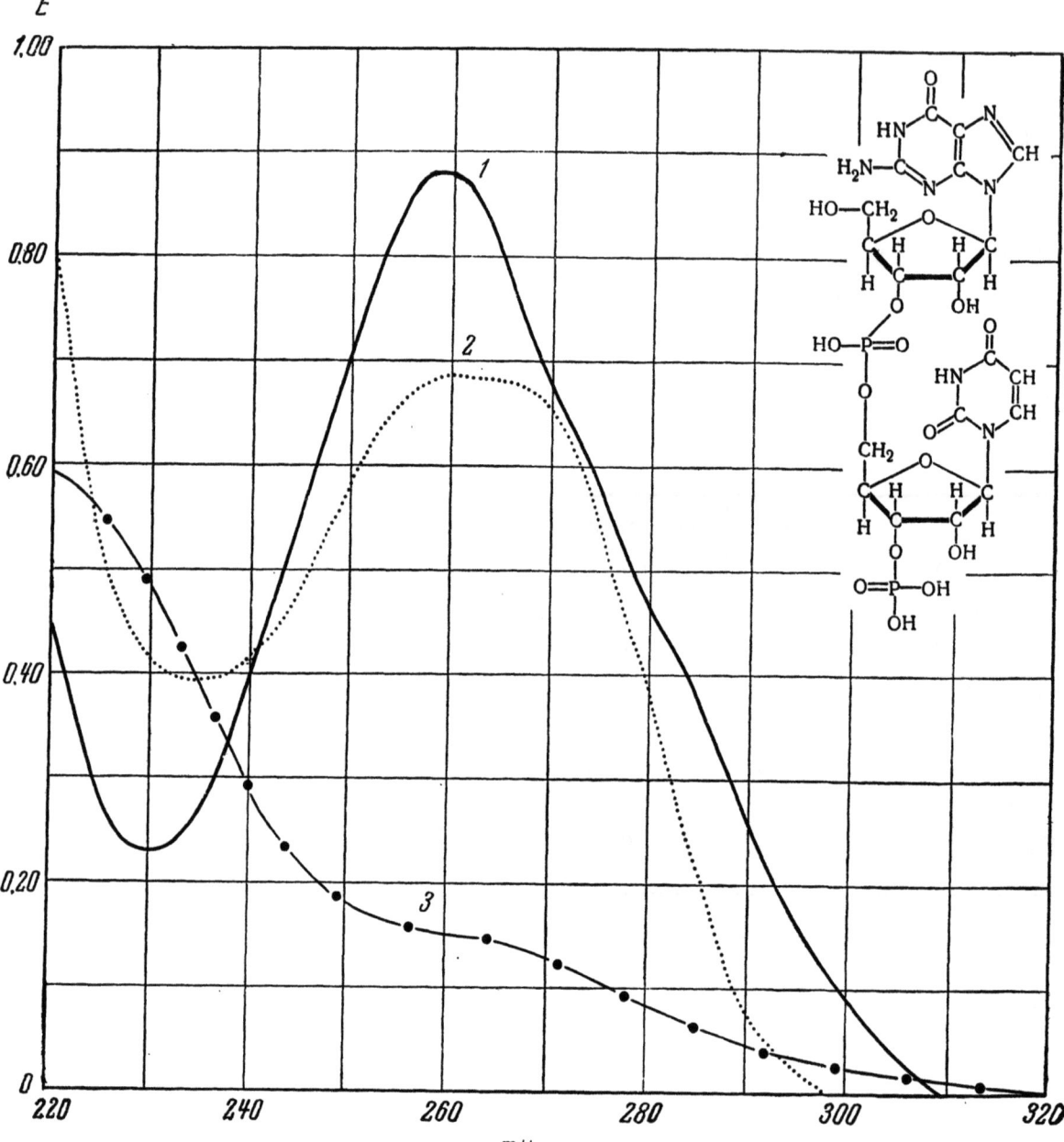

Table 24

GUANYLYL-3',5'-URIDINE-3'-PHOSPHATE [GpUp]

$C_{19}H_{25}N_7O_{16}P_2$

Mol. wt. 669

	0.1 N HCl	0.1 N KOH
λ_{max} (mμ)	259	262
$\varepsilon_{260} \times 10^{-3}$	22.6	—
λ_{min} (mμ)	230	235
$\varepsilon_{min} \times 10^{-3}$	—	—
E_{250}/E_{260}	0.82	0.84
E_{270}/E_{260}	0.77	0.94
E_{280}/E_{260}	0.53	0.56
E_{290}/E_{260}	0.30	0.10

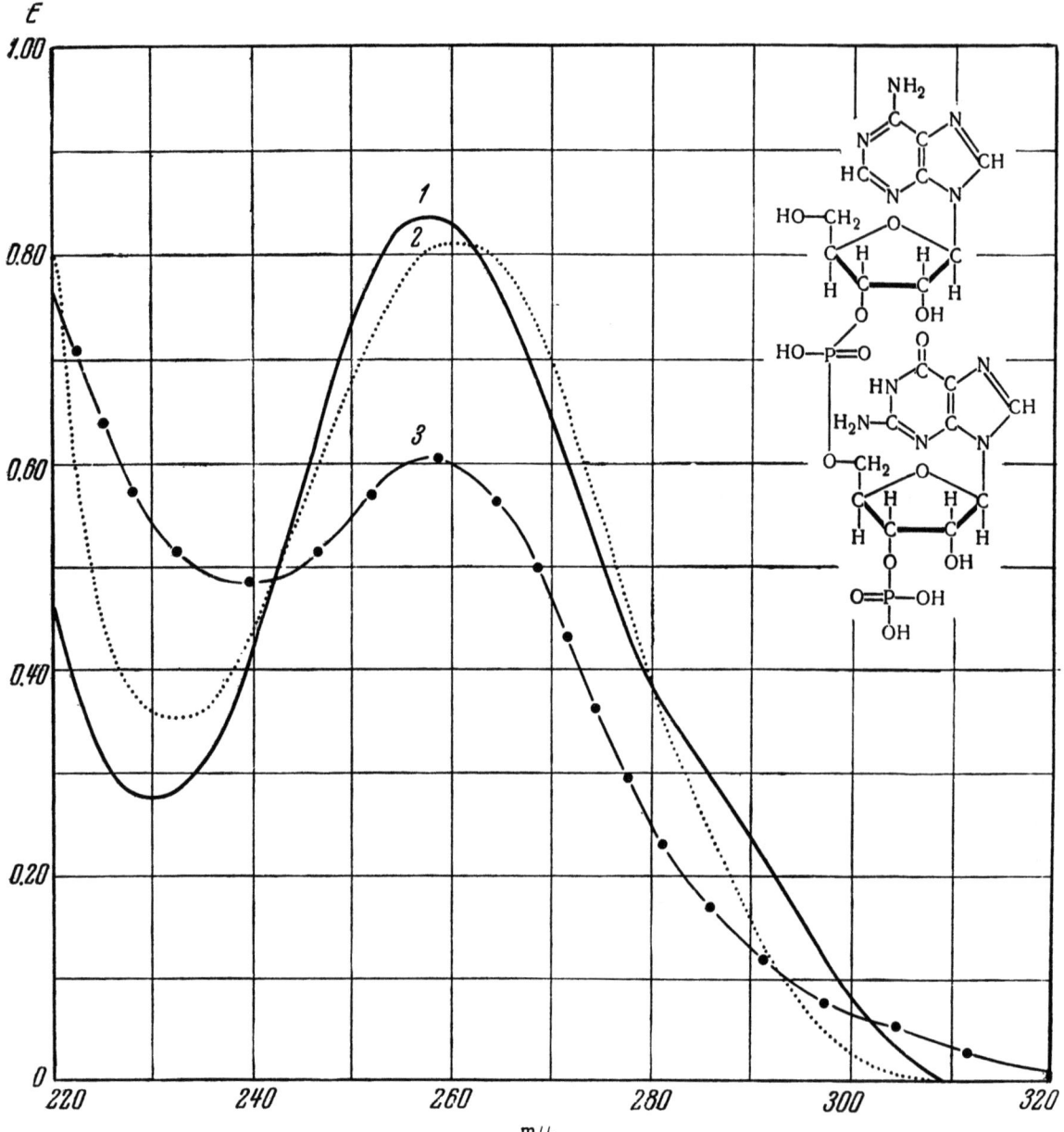

Table 25

ADENYLYL-3',5'-GUANOSINE-3'-PHOSPHATE [ApGp]

$C_{20}H_{26}N_{10}O_{14}P_2$

Mol. wt. 692

	0.1 N HCl	0.1 N KOH
λ_{max} (mμ)	258	260
$\varepsilon_{260} \times 10^{-3}$	—	—
λ_{min} (mμ)	230	232
$\varepsilon_{min} \times 10^{-3}$	—	—
E_{250}/E_{260}	0.88	0.85
E_{270}/E_{260}	0.77	0.85
E_{280}/E_{260}	0.51	0.49
E_{290}/E_{260}	0.35	0.19

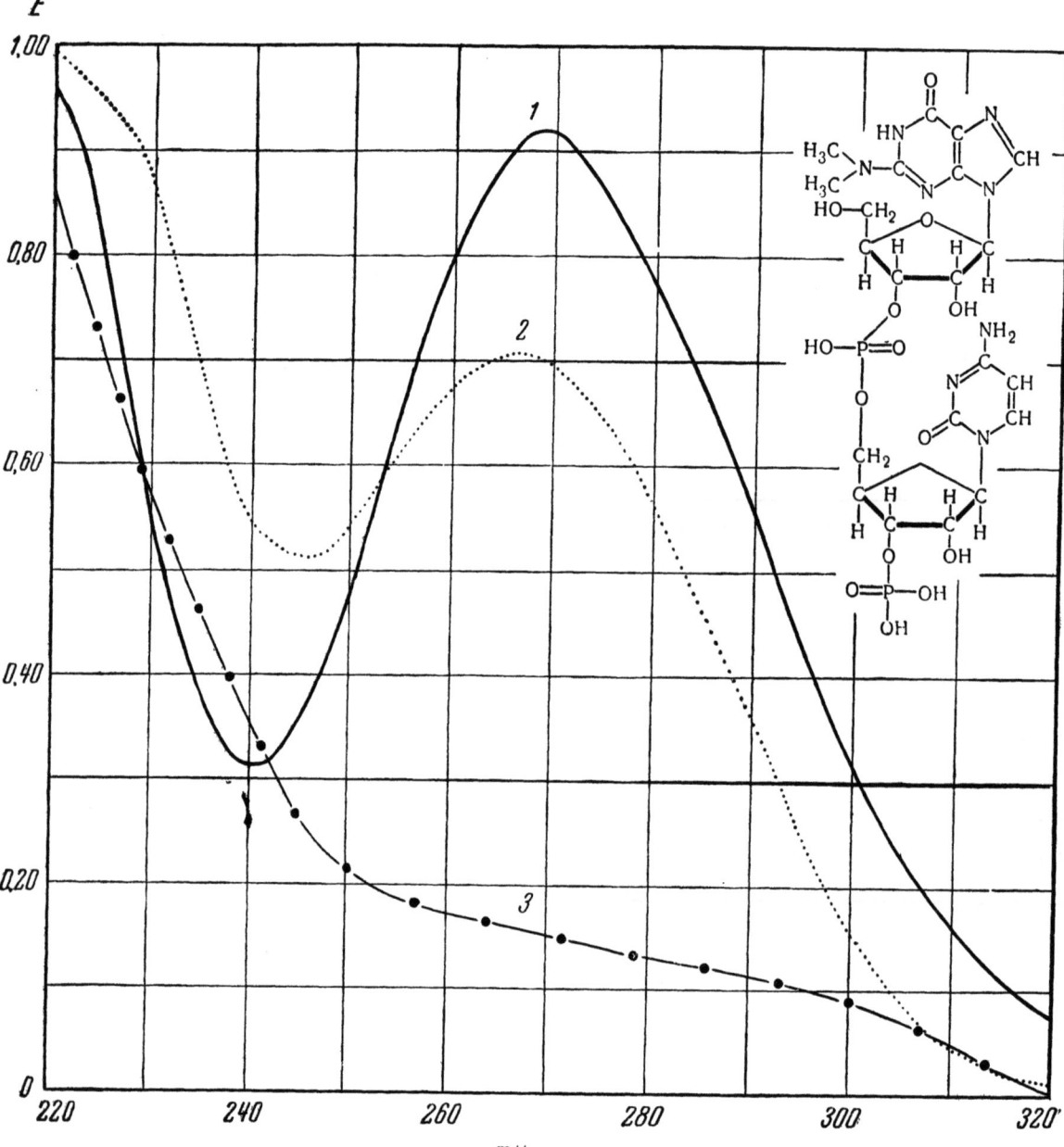

Table 26

N$_2$-DIMETHYLGUANYLYL-3',5'-CYTIDINE-3'-PHOSPHATE

[G^{M}pCp]*

C$_{21}$H$_{30}$N$_8$O$_{15}$P$_2$

Mol. wt. 696

	0.1 N HCl	0.1 N KOH
λ_{max} (mμ)	269	267
$\varepsilon_{260} \times 10^{-3}$	—	—
λ_{min} (mμ)	240	245
$\varepsilon_{min} \times 10^{-3}$	—	—
E_{250}/E_{260}	0.60	0.82
E_{270}/E_{260}	1.13	1.02
E_{280}/E_{260}	0.95	0.85
E_{290}/E_{260}	0.69	0.55

*G^M — N$_2$-dimethylguanosine.

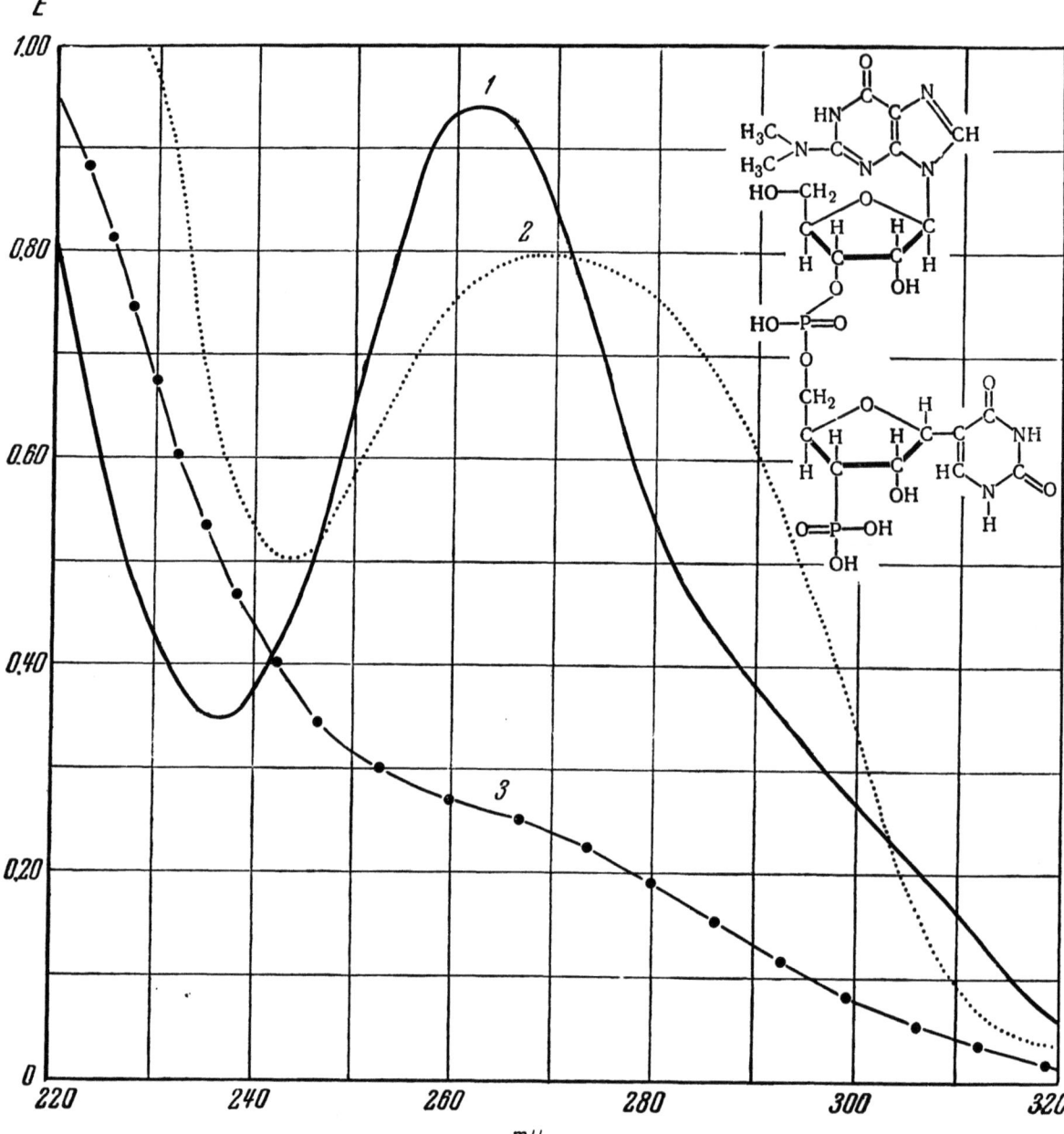

Table 27

N_2-DIMETHYLGUANYLYL-3',5'-PSEUDOURIDINE-3'-PHOSPHATE [$G^M p \psi Up$]*

$C_{21}H_{29}N_7O_{16}P_2$
Mol. wt. 697

	0.1 N HCl	0.1 N KOH
λ_{max} (mμ)	263	268
$\varepsilon_{260} \times 10^{-3}$	23.8	—
λ_{min} (mμ)	236	243
$\varepsilon_{min} \times 10^{-3}$	—	—
E_{250}/E_{260}	0.70	0.78
E_{270}/E_{260}	0.91	1.04
E_{280}/E_{260}	0.57	0.99
E_{290}/E_{260}	0.41	0.80

*G^M — N_2-dimethylguanosine.

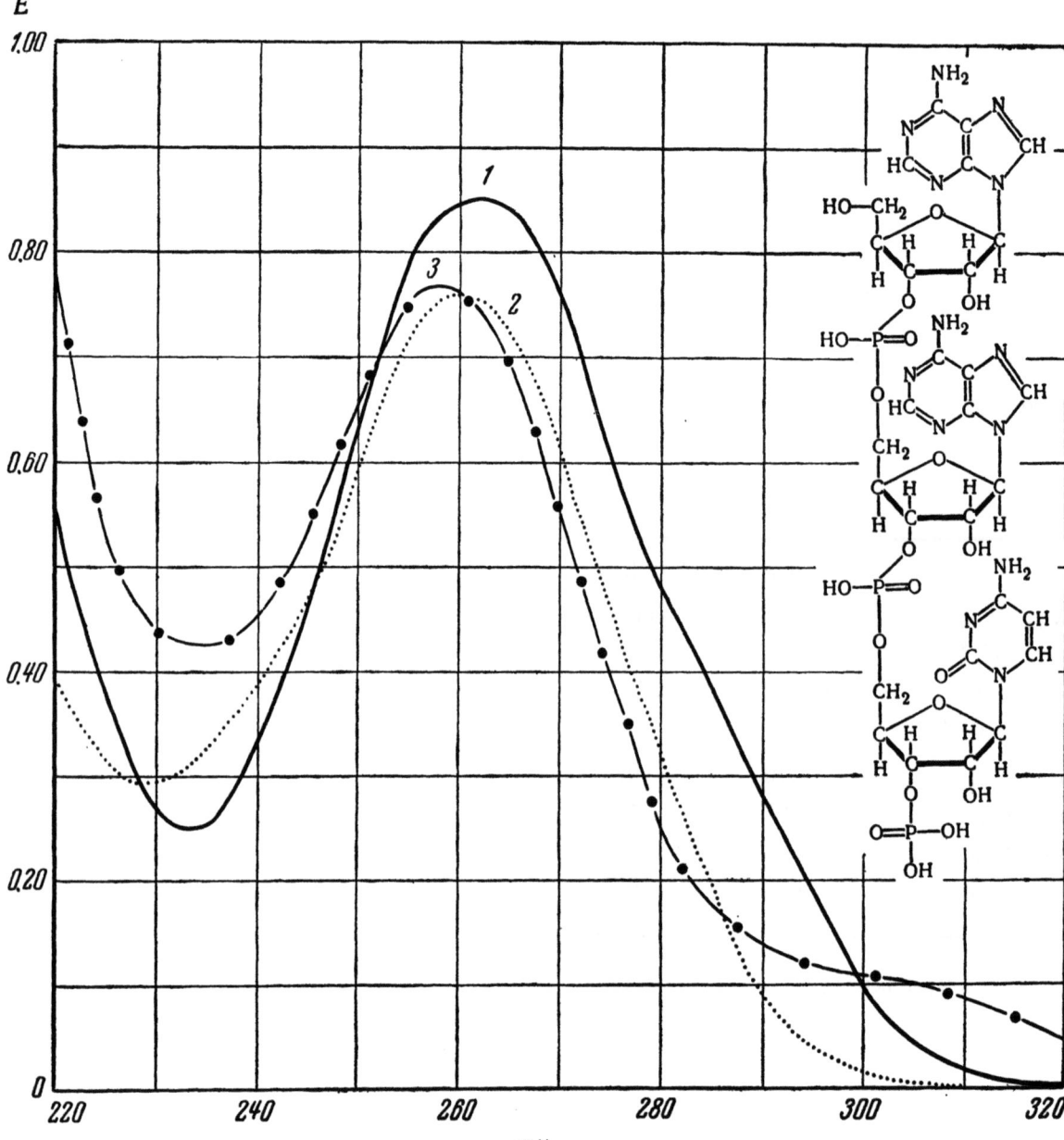

Table 28

ADENYLYL-3',5'-ADENYLYL-3',5'-CYTIDINE-3'-PHOSPHATE

[ApApCp]

$C_{29}H_{36}N_{13}O_{20}P_3$

Mol. wt. 982

	0.1 N HCl	0.1 N KOH
λ_{max} (mμ)	262	260
$\varepsilon_{260} \times 10^{-3}$	—	—
λ_{min} (mμ)	233	230
$\varepsilon_{min} \times 10^{-3}$	—	—
E_{250}/E_{260}	0.75	0.79
E_{270}/E_{260}	0.90	0.80
E_{280}/E_{260}	0.57	0.41
E_{290}/E_{260}	0.33	0.12

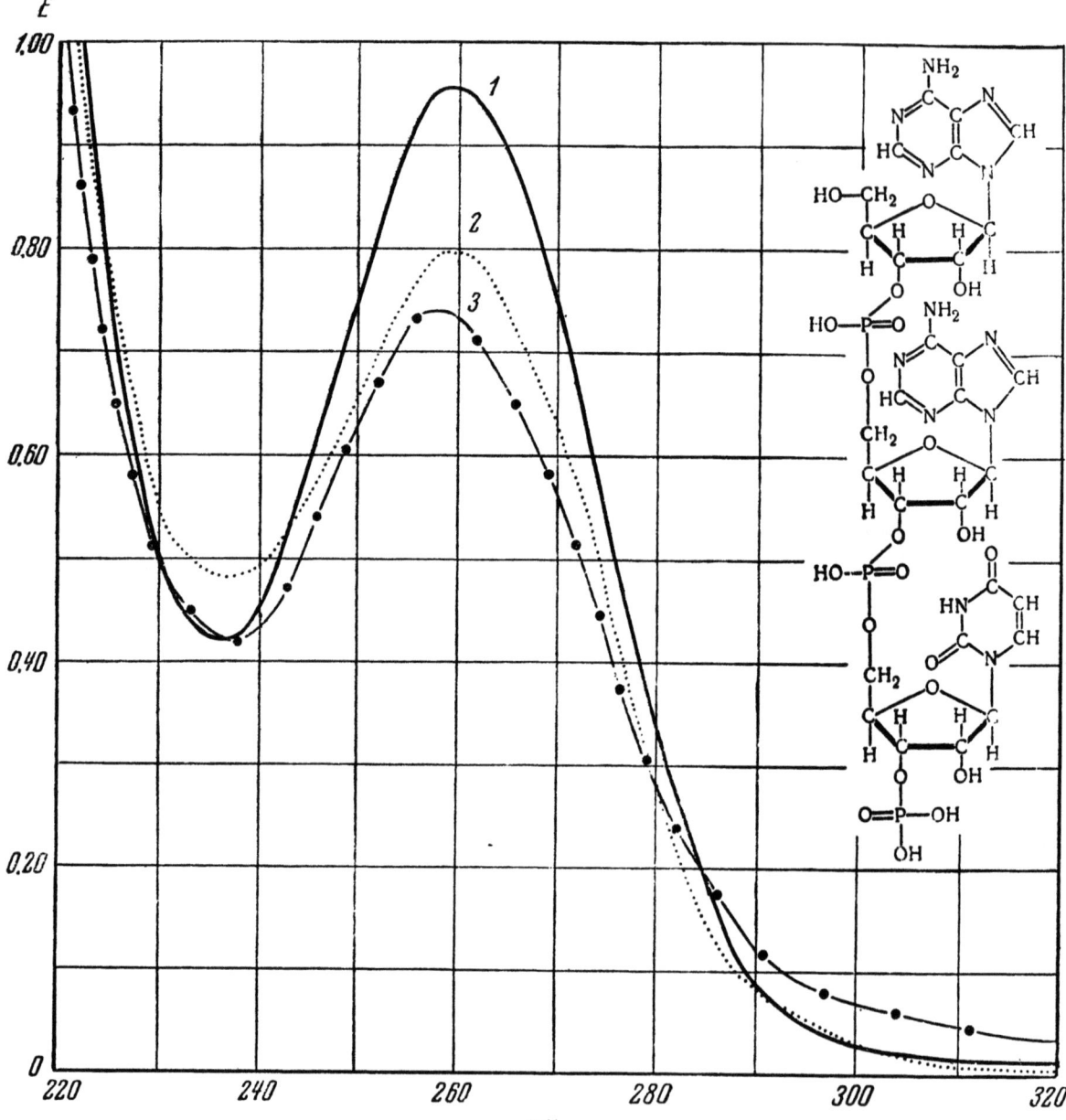

Table 29

ADENYLYL-3', 5'-ADENYLYL-3', 5'-URIDINE-3'-PHOSPHATE

[ApApUp]

$C_{29}H_{37}N_{12}O_{21}P_3$

Mol. wt. 983

	0.1 N HCl	0.1 N KOH
λ_{max} (mμ)	260	260
$\varepsilon_{260} \times 10^{-3}$	—	—
λ_{min} (mμ)	235	235
$\varepsilon_{min} \times 10^{-3}$	—	—
E_{250}/E_{260}	0.79	0.83
E_{270}/E_{260}	0.79	0.78
E_{280}/E_{260}	0.35	0.35
E_{290}/E_{260}	0.09	0.10

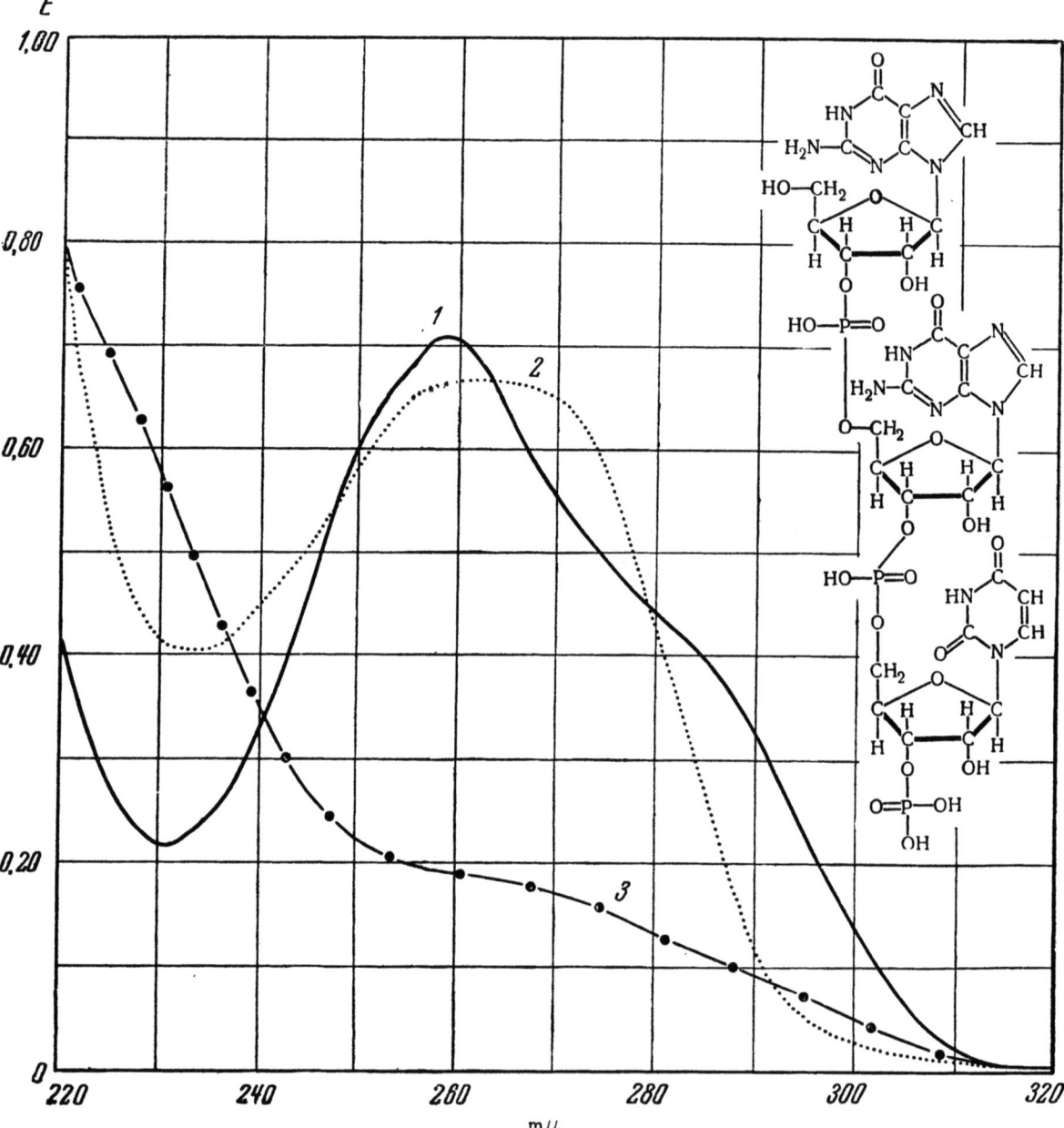

Table 30

GUANYLYL-3',5'-GUANYLYL-3',5'-URIDINE-3'-PHOSPHATE
[GpGpUp]
$C_{29}H_{37}N_{12}O_{23}P_3$
Mol. wt. 1014

	0.1 N HCl	0.1 N KOH
λ_{max} (mμ)	259	263—266
$\varepsilon_{260} \times 10^{-3}$	—	—
λ_{min} (mμ)	230	233
$\varepsilon_{min} \times 10^{-3}$	—	—
E_{250}/E_{260}	0.88	0.86
E_{270}/E_{260}	0.80	0.95
E_{280}/E_{260}	0.60	0.60
E_{290}/E_{260}	0.45	0.15

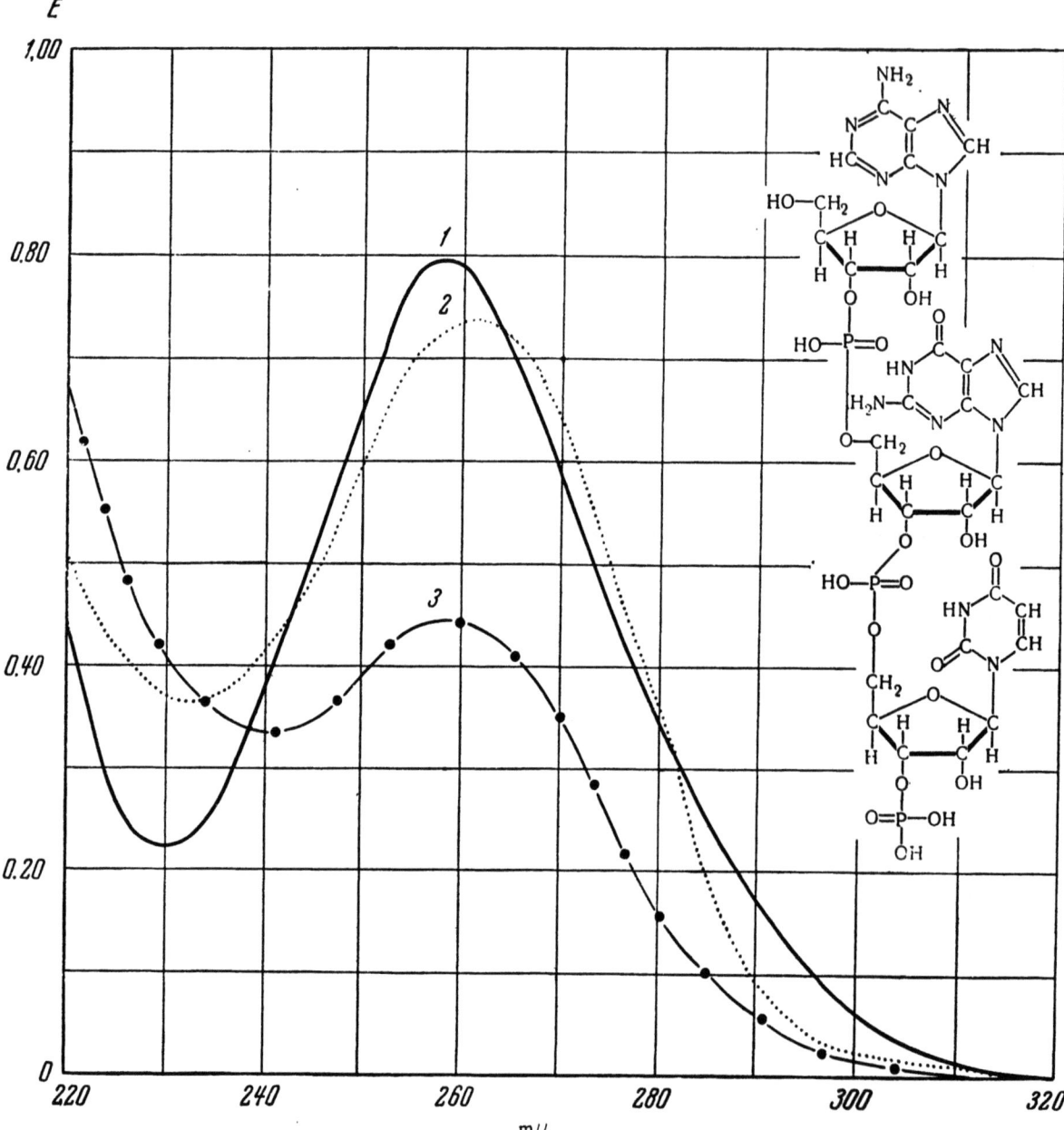

Table 31

(ADENYLYL-3',5'-GUANYLYL-3',5')-URIDINE-3'-PHOSPHATE [(ApGp)Up]*

$C_{29}H_{37}N_{12}O_{22}P_3$

Mol. wt. 998

	0.1 N HCl	0.1 N KOH
λ_{max} (mμ)	258	262
$\varepsilon_{260} \times 10^{-3}$	—	—
λ_{min} (mμ)	230	233
$\varepsilon_{min} \times 10^{-3}$	—	—
E_{250}/E_{260}	0.84	0.83
E_{270}/E_{260}	0.77	0.85
E_{280}/E_{260}	0.42	0.45
E_{290}/E_{260}	0.20	0.10

*Mixture of ApGpUp and GpApUp.

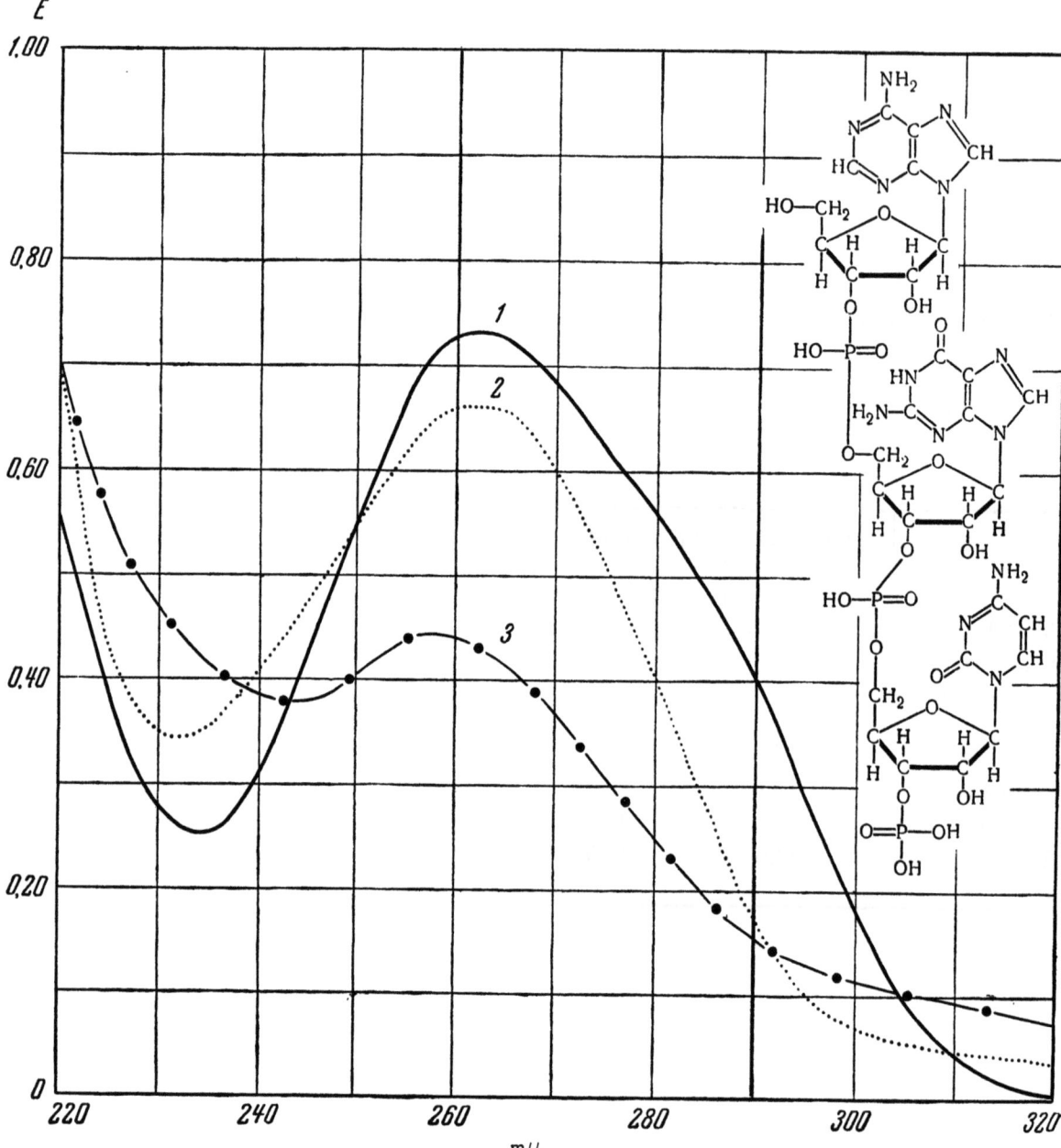

Table 32

(ADENYLYL-3',5'-GUANYLYL-3',5')-CYTIDINE-3'-PHOSPHATE [(ApGp)Cp]*

$C_{29}H_{38}N_{13}O_{21}P_3$

Mol. wt. 998

	0.1 N HCl	0.1 N KOH
λ_{max} (mμ)	263	263
$\varepsilon_{260} \times 10^{-3}$	—	—
λ_{min} (mμ)	234	232
$\varepsilon_{min} \times 10^{-3}$	—	—
E_{250}/E_{260}	0.75	0.85
E_{270}/E_{260}	0.95	0.92
E_{280}/E_{260}	0.77	0.62
E_{290}/E_{260}	0.55	0.25

*Mixture of ApGpCp and GpApCp.

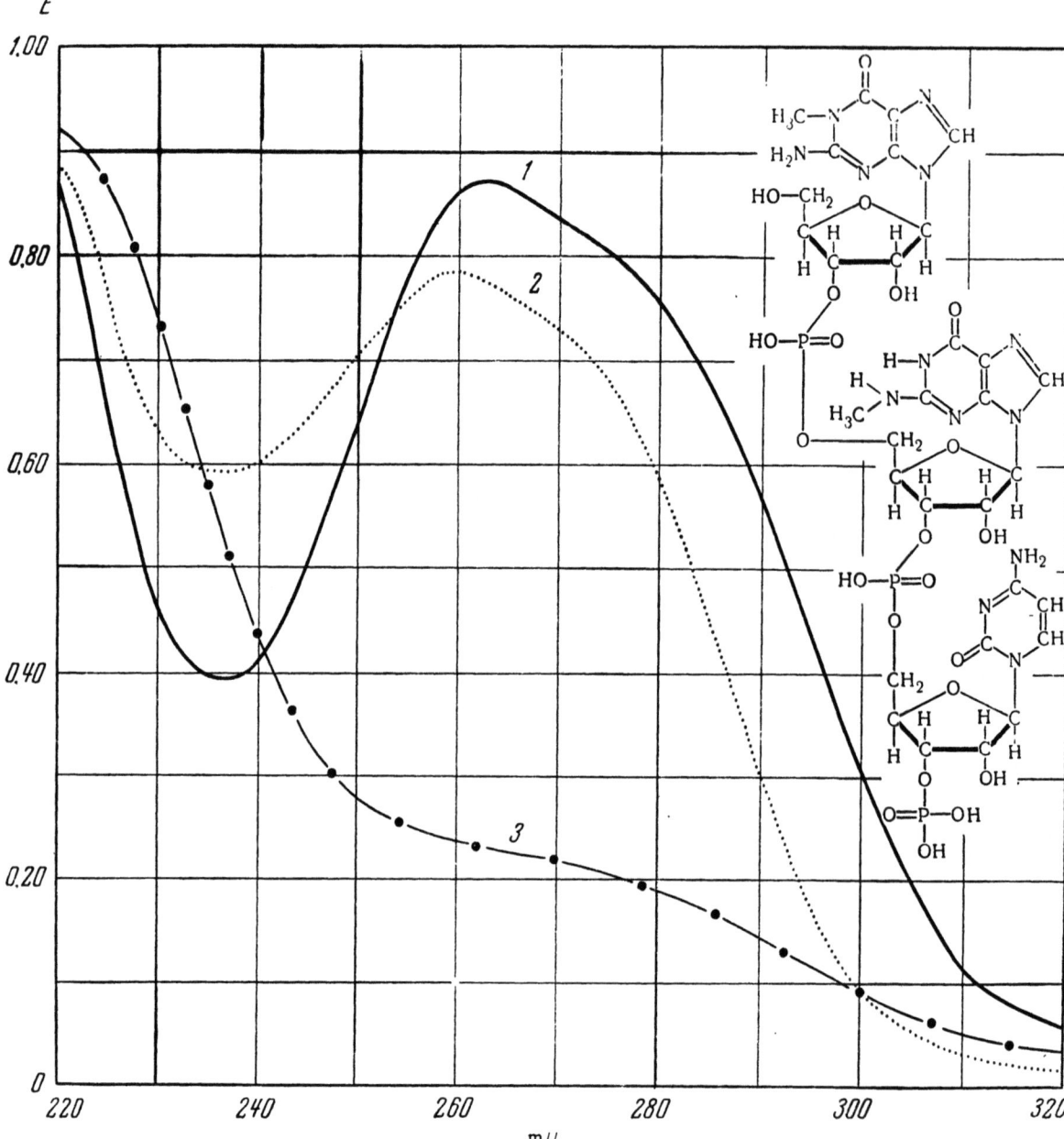

Table 33

**1-METHYLGUANYLYL-
3',5'-N$_2$-METHYLGUANYLYL-
3',5'-CYTIDINE-3'-PHOSPHATE**
[G^{M_1}pG^{M_2}pCp]*
C$_{31}$H$_{42}$N$_{13}$O$_{22}$P$_3$
Mol. wt. 1042

	0.1 N HCl	0.1 N KOH
λ_{max} (mμ)	263	260
$\varepsilon_{260} \times 10^{-3}$	—	—
λ_{min} (mμ)	237	236
$\varepsilon_{min} \times 10^{-3}$	—	—
E_{250}/E_{260}	0.75	0.90
E_{270}/E_{260}	0.92	0.94
E_{280}/E_{260}	0.87	0.75
E_{290}/E_{260}	0.66	0.38

*G^{M_1} — 1-methylguanosine; G^{M_2} — N$_2$-methylguanosine.

Literature Cited

Introduction
1. T. V. Venkstern, A. D. Mirzabekov, V. I. Gorshkova, and A. A. Baev, Biokhimiya, 28:712 (1963).
2. T. V. Venkstern, A. A. Baev, A. D. Mirzabekov, and V. I. Gorshkova, Dokl. Akad. Nauk SSSR, 151:220 (1963).
3. E. D. Dovedova, Biokhimiya, 24:414 (1959).
4. A. I. Krutilina, T. V. Venkstern, and A. A. Baev, Biokhimiya, 29:333 (1964).
5. M. F. Khanina, T. V. Venkstern, and A. A. Baev, Biokhimiya, 29:142 (1964).
6. R. Bergquist and A. Deutsch, Acta Chem. Scand., 8:1880 (1954).
7. G. L. Cantoni, H. V. Gelboin, S. W. Luborsky, and H. H. Richards, Biochim. Biophys. Acta, 61:354 (1962).
8. R. Caputto, L. F. Leloir, C. E. Cardini, and A. C. Paladini, J. Biol. Chem., 184:333 (1950).
9. D. B. Dunn and J. D. Smith, Nature, 175:336 (1955).
10. D. B. Dunn and J. D. Smith, Biochem. J., 68:627 (1958).
11. D. B. Dunn and J. D. Smith, IV Intern. Congr. Biochem. (Vienna, 1958), Vol. 7, Pergamon Press (1959).
12. D. B. Dunn, J. H. Hitchborn, and A. R. Trim, Biochem. J., 88:34P (1963).
13. G. B. Elion, W. H. Lange, and G. H. Hitchings, J. Am. Chem. Soc., 78:217 (1956).
14. D. Hamer, D. M. Waldron and D. L. Woodhouse, Arch. Biochem. Biophys., 47:272 (1953).
15. A. S. Jones and D. L. Woodhouse, Nature, 183:1603 (1959).
16. B. Magasanik, E. Visher, R. Doniger, D. Elson, and E. Chargaff, J. Biol. Chem., 186:37 (1950).
17. G. W. Rushizky and C. A. Knight, Virology, 11:236 (1960).
18. H. S. Shapiro and E. Chargaff, Biochem. Biophys. Acta, 39:62 (1960).
19. J. D. Smith and D. B. Dunn, Biochem. J., 72:294 (1959).
20. M. Soodak, A. Pircio, and L. R. Cerecedo, J. Biol. Chem., 181:713 (1949).
21. T. Suzuki and E. Ito, J. Biochem. (Tokyo), 45:403 (1958).
22. G. R. Wyatt, Nature, 166:237 (1950).
23. G. R. Wyatt, Biochem. J., 48:581 (1951).
24. G. R. Wyatt and S. S. Cohen, Nature, 170:1072 (1952).
25. C. T. Yu and P. C. Zamecnik, Biochim. Biophys. Acta, 76:209 (1963).

Methylated Derivatives of Adenine
1. G. N. Zaitseva, T. M. Dmitrieva, Hsü Ch'ang-fa, and A. N. Belozerskii, Dokl. Akad. Nauk SSSR, 147:1211 (1962).

2. M. Adler, B. Weissmann, A. B. Gutman, J. Biol. Chem., 230:717 (1958).
3. P. Brookes and P. D. Lawley, J. Chem. Soc., 539 (1960).
4. B. E. Griffin and C. B. Reese, Biochim. Biophys. Acta, 68:185 (1963).
5. D. B. Dunn, Biochim. Biophys. Acta, 46:198 (1961).
6. D. B. Dunn, Biochem. J., 86:14P (1963).
7. D. B. Dunn and J. D. Smith, Nature, 175:336 (1955).
8. D. B. Dunn and J. D. Smith, Biochem. J., 68:627 (1958).
9. D. B. Dunn, J. D. Smith, and P. F. Spahr, J. Mol. Biol., 2:113 (1960).
10. G. Elion, E. Burgi, and G. H. Hitchings, J. Am. Chem. Soc., 74:411 (1952).
11. R. H. Hall, Biochim. Biophys. Acta, 68:278 (1963).
12. J. W. Littlefield and D. B. Dunn, Nature, 181:254 (1958).
13. J. W. Littlefield and D. B. Dunn, Biochem. J., 68:8P (1958).
14. J. W. Littlefield and D. B. Dunn, Biochem. J., 70:642 (1958).
15. B. Weissmann, P. A. Bromberg, A. B. Gutman, J. Biol. Chem., 224:407 (1957).

Methylated Derivatives of Guanine
1. G. N. Zaitseva, T. M. Dmitrieva, Hsü Ch'ang-fa, and A. N. Belozerskii, Dokl. Akad. Nauk SSSR, 147:1211 (1962).
2. M. Adler, B. Weissmann, and A. B. Gutman, J. Biol. Chem., 230:717 (1958).
3. J. M. Gulland and L. F. Story, J. Chem. Soc., 692 (1938).
4. F. F. Davis, A. F. Carlucci, and J. F. Roubein, J. Biol. Chem., 234:1525 (1959).
5. D. B. Dunn, Abstr. Sympos. Ribonucleic Acids and Polyphosphates (Strasbourg, July, 1961).
6. D. B. Dunn, Biochem. J., 86:14P (1963).
7. D. B. Dunn and J. D. Smith, Trans. Faraday Soc., 59:490 (1959).
8. D. B. Dunn and J. D. Smith, IV Intern. Congr. Biochem. (Vienna, 1958), Vol. 7, Pergamon Press (1959).
9. W. F. Hemmens, Biochim. Biophys. Acta, 68:284 (1963).
10. J. W. Kemp and F. W. Allen, Biochim. Biophys. Acta, 28:51 (1958).
11. P. N. Magee and E. Farber, Biochem. J., 83:114 (1962).
12. B. Reiner and S. Zamenhof, J. Biol. Chem., 228:475 (1957).
13. J. D. Smith and D. B. Dunn, Biochem. J., 72:294 (1959).
14. S. Udenfriend and P. Saltzman, Anal. Biochem., 3:49 (1962).
15. S. Udenfriend, P. Saltzman-Nirenberg, G. L. Cantoni, Anal. Biochem., 5:258 (1963).
16. B. Weissmann, P. A. Bromberg, and A. B. Gutman, J. Biol. Chem., 224:423 (1957).

Pseudouridine and Pseudouridylic Acid
1. N. A. Gumilevskaya and N. M. Sisakyan, Dokl. Akad. Nauk SSSR, 144:223 (1962).
2. G. N. Zaitseva, T. M. Dmitrieva, Hsü Ch'ang-fa, and A. N. Belozerskii, Dokl. Akad. Nauk SSSR, 147:1211 (1962).
3. W. S. Adams, F. Davis, and M. Nakatami, Am. J. Med., 28:726 (1960).
4. M. Adler and A. B. Gutman, Science, 130:862 (1959).
5. G. Brawerman and E. Chargaff, Biochim. Biophys. Acta, 31:172 (1959).
6. G. Brawerman, D. A. Hufnagel, and E. Chargaff, Biochim. Biophys. Acta, 61:340 (1962).
7. R. W. Chambers, V. Kurkow, and R. Shapiro, Biochemistry, 2:1192 (1963).
8. W. E. Cohn, J. Am. Chem. Soc., 72:1471 (1950).
9. W. E. Cohn and E. Volkin, Nature, 167:483 (1951).
10. W. E. Cohn, Federation Proc., 16:166 (1957).
11. W. E. Cohn, Federation Proc., 17:203 (1958).
12. W. E. Cohn, Biochim. Biophys. Acta, 32:569 (1959).

13. W. E. Cohn, J. Biol. Chem., 235:1488 (1960).
14. A. M. Crestfield and F. W. Allen, Anal. Chem., 27:422 (1955).
15. F. F. Davis and F. W. Allen, J. Biol. Chem., 227:907 (1957).
16. F. F. Davis, A. F. Carlucci, and J. F. Roubein, J. Biol. Chem., 234:1525 (1959).
17. D. B. Dunn, Biochim. Biophys. Acta, 34:286 (1959).
18. D. B. Dunn, J. D. Smith, and P. F. Spahr, J. Mol. Biol., 2:113 (1960).
19. A. Dlugajczyk and F. W. Allen, Biochim. Biophys. Acta, 51:215 (1961).
20. A. Dlugajczyk and J. J. Eiler, Federation Proc., 22:470 (1963).
21. K. Fink and W. S. Adams, Federation Proc., 22:470 (1963).
22. D. G. Glitz and Ch. A. Dekker, Biochemistry, 2:1185 (1963).
23. J. H. Goldberg and M. Rabinowitz, Biochim. Biophys. Acta, 54:202 (1961).
24. J. W. Kemp and F. W. Allen, Biochim. Biophys. Acta, 28:51 (1958).
25. B. G. Lane and F. W. Allen, Biochim. Biophys. Acta, 47:36 (1961).
26. R. Lipshitz and E. Chargaff, Biochim. Biophys. Acta, 42:544 (1960).
27. A. M. Michelson and W. E. Cohn, Biochemistry, 1:490 (1962).
28. S. Osawa, Biochim. Biophys. Acta, 42:244 (1960).
29. E. Otaka, J. Hotta, and S. Osawa, Biochim. Biophys. Acta, 35:266 (1959).
30. R. Shapiro and R. W. Chambers, J. Am. Chem. Soc., 83:3920 (1961).
31. M. Tada, J. Biochem. (Tokyo), 51:92 (1962).
32. C. T. Yu and F. W. Allen, Biochim. Biophys. Acta, 32:393 (1959).

Methylated Derivatives of Uracil
1. G. N. Zaitseva, T. M. Dmitrieva, Hsü Ch'ang-fa, and A. N. Belozerskii, Dokl. Akad. Nauk SSSR, 147:1211 (1962).
2. P. Berg, Ann. Rev. Biochem., 30:293 (1961).
3. G. L. Cantoni, Abstr. Sympos. Ribonucleic Acids and Polyphosphates (Strasbourg, July, 1961).
4. F. F. Davis, A. F. Carlucci, and J. F. Roubein, J. Biol. Chem., 234:1525 (1959).
5. D. B. Dunn, Biochim. Biophys. Acta, 34:286 (1959).
6. D. B. Dunn, Biochim. Biophys. Acta, 38:176 (1960).
7. D. B. Dunn, J. D. Smith, P. F. Spahr, J. Mol. Biol. 2:113 (1960).
8. J. J. Fox and D. Shugar, Biochim. Biophys. Acta, 9:369 (1952).
9. J. J. Fox, N. Jung, J. Davoll, and G. B. Brown, J. Am. Chem. Soc., 78:2117 (1956).
10. R. H. Hall, Biochem. Biophys. Res. Commun., 12:361 (1963).
11. J. W. Littlefield and D. B. Dunn, Biochem. J., 70:642 (1958).
12. J. W. Littlefield and D. B. Dunn, Nature, 181:254 (1958).
13. J. W. Littlefield and D. B. Dunn, Biochem. J., 68:8P (1958).
14. R. Monier, M. L. Stephenson, and P. C. Zamecnik, Biochim. Biophys. Acta, 43:1 (1960).
15. T. D. Price, H. A. Hinds, and R. S. Brown, Federation Proc., 21:376 (1962).
16. T. D. Price, H. A. Hinds, and R. S. Brown, J. Biol. Chem., 238:311 (1963).

Methylated Derivatives of Cytosine
1. G. N. Zaitseva, T. M. Dmitrieva, Hsü Ch'ang-fa, and A. N. Belozerskii, Dokl. Akad. Nauk SSSR, 147:1211 (1962).
2. H. Amos and M. Korn, Biochim. Biophys. Acta, 29:444 (1958).
3. P. Brooks and P. D. Lawley, J. Chem. Soc., 1348 (1962).
4. G. Brawerman and E. Chargaff, Biochim. Biophys. Acta, 31:172 (1959).
5. G. L. Cantoni, Abstr. Sympos. Ribonucleic Acids and Polyphosphates (Strasbourg, July, 1961).
6. G. L. Cantoni, H. V. Gelboin, S. W. Luborsky, and H. H. Richards, Biochim. Biophys. Acta, 61:354 (1962).

7. D. B. Dunn, Biochim. Biophys. Acta, 34:286 (1959).
8. D. B. Dunn and J. D. Smith, IV Intern. Congr. Biochem. (Vienna, 1958), Vol. 7, Pergamon Press (1959).
9. D. B. Dunn, Biochim. Biophys. Acta, 38:176 (1960).
10. D. B. Dunn, J. D. Smith, P. F. Spahr, J. Mol. Biol., 2:113 (1960).
11. J. J. Fox and D. Shugar, Biochim. Biophys. Acta, 9:369 (1952).
12. J. J. Fox, D. Van Praag, J. Wempen, G. L. Doerr, L. Cheong, J. E. Knoll, M. L. Edinoff, A. Bendich, and G. B. Brown, J. Am. Chem. Soc., 81:178 (1959).
13. R. H. Hall, Biochem. Biophys. Res. Commun., 12:361 (1963).
14. S. G. Laland, W. G. Overend, and M. Webb, J. Chem. Soc., 3224 (1952).
15. T. D. Price, H. A. Hinds, R. S. Brown, J. Biol. Chem., 238:311 (1963).
16. M. Sluyser and L. Bosch, Biochim. Biophys. Acta, 55:479 (1962).
17. G. R. Wyatt, Biochem. J., 48:581 (1951).
18. G. R. Wyatt, Biochem. J., 48:584 (1951).
19. G. R. Wyatt and S. S. Cohen, Nature, 170:1072 (1952).
20. G. R. Wyatt and S. S. Cohen, Biochem. J., 55:774 (1953).

Oligonucleotides
1. T. V. Venkstern, A. D. Mirzabekov, V. I. Gorshkova, and A. A. Baev, Biokhimiya, 28:712 (1963).
2. T. V. Venkstern, A. A. Baev, A. D. Mirzabekov, and V. I. Gorshkova, Dokl. Akad. Nauk SSSR, 151:220 (1963).
3. V. M. Ingram and J. G. Pierce, Biochemistry, 1:580 (1962).
4. J. T. Madison, G. A. Everett, A. W. Holley, and Ada Zamir, Federation Proc., 22:230 (1963).
5. G. W. Rushizky, Federation Proc., 21:371 (1962).
6. G. W. Rushizky and C. A. Knight, Biochem. Biophys. Res. Commun., 2:66 (1960).
7. G. W. Rushizky and C. A. Knight, Virology, 11:236 (1960).
8. M. Staehelin, Biochim. Biophys. Acta, 49:11 (1961).
9. M. Staehelin, Biochem. J., 89:2P (1963).
10. M. Staehelin, H. G. Zachau, and M. Schweiger, Biochem. J., 84:107P (1962).
11. M. Staehelin, M. Schweiger, and H. G. Zachau, Biochim. Biophys. Acta, 68:129 (1963).

MIX
Papier aus verantwortungsvollen Quellen
Paper from responsible sources
FSC® C105338

If you have any concerns about our products,
you can contact us on
ProductSafety@springernature.com

In case Publisher is established outside the EU,
the EU authorized representative is:
**Springer Nature Customer Service Center GmbH
Europaplatz 3, 69115 Heidelberg, Germany**

Printed by Libri Plureos GmbH
in Hamburg, Germany